AF458779

ON THE GRID

The Berkeley Tanner Lectures

The Tanner Lectures on Human Values were established by the American scholar, industrialist, and philanthropist Obert Clark Tanner; they are presented annually at nine universities in the United States and England. The University of California, Berkeley, became a permanent host of annual Tanner Lectures in the academic year 2000–2001. This work is the seventeenth in a series of books based on the Berkeley Tanner Lectures. The volume includes a revised version of the lectures that Michael Warner presented at Berkeley in March 2018, together with commentaries by Dale Jamieson, Anahid Nersessian, and Jedediah Britton-Purdy. The volume is edited by Michael Lucey, who also contributes an introduction.

The Berkeley Tanner Lecture Series was established in the belief that these distinguished lectures, together with the lively debates stimulated by their presentation in Berkeley, deserve to be made available to a wider audience. Additional volumes are in preparation.

Kinch Hoekstra
R. Jay Wallace
Series Editors

Selected Volumes in the Series

Frank Kermode, *Pleasure and Change: The Aesthetics of Canon*
Edited by Robert Alter
With Geoffrey Hartman, John Guillory, and Carey Perloff

Seyla Benhabib, *Another Cosmopolitanism*
Edited by Robert Post
With Jeremy Waldron, Bonnie Honig, and Will Kymlicka

Axel Honneth, *Reification: A New Look at an Old Idea*
Edited by Martin Jay
With Judith Butler, Raymond Guess, and Jonathan Lear

Samuel Scheffler, *Death and the Afterlife*
Edited by Niko Kolodny
With Susan Wolf, Harry G. Frankfurt, and Seana Valentine Shiffrin

Eric Santner, *The Weight of All Flesh: On the Subject Matter of Political Economy*
Edited by Kevis Goodman
With Bonnie Honig, Peter E. Gordon, and Hent de Vries

Didier Fassin, *The Will to Punish*
Edited by Christopher Kutz
With Bruce Western, Rebecca M. McLennan, and David W. Garland

Philip Pettit, *The Birth of Ethics*
Edited by Kinch Hoekstra
With Michael Tomasello

ON THE GRID

Climate Change and the Utopia of Green Energy

Michael Warner

With Commentaries by

Dale Jamieson

Jedediah Britton-Purdy

Anahid Nersessian

Edited and Introduced by

Michael Lucey

OXFORD
UNIVERSITY PRESS

Oxford University Press is a department of the University of Oxford. It furthers the University's objective of excellence in research, scholarship, and education by publishing worldwide. Oxford is a registered trade mark of Oxford University Press in the UK and certain other countries.

Published in the United States of America by Oxford University Press
198 Madison Avenue, New York, NY 10016, United States of America.

© Regents of the University of California 2025

All rights reserved. No part of this publication may be reproduced, stored in a retrieval system, transmitted, used for text and data mining, or used for training artificial intelligence, in any form or by any means, without the prior permission in writing of Oxford University Press, or as expressly permitted by law, by license or under terms agreed with the appropriate reprographics rights organization. Inquiries concerning reproduction outside the scope of the above should be sent to the Rights Department, Oxford University Press, at the address above.

You must not circulate this work in any other form
and you must impose this same condition on any acquirer.

Library of Congress Cataloging-in-Publication Data
Names: Warner, Michael, 1958– author
Title: On the grid : climate change and the utopia of green energy / Michael Warner.
Description: New York : Oxford University Press, 2025. |
Series: The Berkeley Tanner lectures |
Identifiers: LCCN 2025025514 (print) | LCCN 2025025515 (ebook) |
ISBN 9780197696248 hardback | ISBN 9780197696279 electronic |
ISBN 9780197696262 epub
Subjects: LCSH: Renewable energy sources | Climate change mitigation |
Environmental protection—Citizen participation | Energy consumption |
Energy policy
Classification: LCC TJ808 .W365 2025 (print) | LCC TJ808 (ebook)
LC record available at https://lccn.loc.gov/2025025514
LC ebook record available at https://lccn.loc.gov/2025025515

DOI: 10.1093/oso/9780197696248.001.0001

Printed by Integrated Books International, United States of America

The manufacturer's authorized representative in the EU for product safety is Oxford University Press España S.A., Parque Empresarial San Fernando de Henares, Avenida de Castilla, 2 – 28830 Madrid (www.oup.es/en or product.safety@oup.com). OUP España S.A. also acts as importer into Spain of products made by the manufacturer.

CONTENTS

COMMENTARIES

RESPONSE

ACKNOWLEDGMENTS

Grateful acknowledgment is due to all those who made the Tanner Lectures at Berkeley happen—particularly to Martin Jay for his inspiration, Jay Wallace for his patience and grace, and Michael Lucey for everything. The friends and colleagues who gave me help are too many to name, so I trust they will know how much I appreciate them. Caleb Smith and John Durham Peters gave me especially useful comments. Dipesh Chakrabarty opened some of the book's central issues and remains an inspiration. I am also grateful to the students in my classes who have hashed out with me various parts of the book; they were an invaluable gauge of what matters. I am grateful most of all to the respondents who brought such intelligence and goodwill to the conversation, as any reader of their comments will appreciate.

CONTRIBUTORS

Michael Warner is Seymour H. Knox Professor of English at Yale University. He received his PhD from Johns Hopkins and taught at Northwestern and Rutgers before going to Yale, where he served as chair of the Department of English. His books include *Publics and Counterpublics* (Zone Books, 2002); *The Trouble with Normal* (Free Press, 1999); and *The Letters of the Republic: Publication and the Public Sphere in Eighteenth-Century America* (Harvard University Press, 1990). With Craig Calhoun and Jonathan VanAntwerpen, he has edited *Varieties of Secularism in a Secular Age* (Harvard University Press, 2010). He is also the editor of *The Portable Walt Whitman* (Penguin, 2003); *American Sermons* (Library of America, 1999); *The English Literatures of America, 1500–1800* (with Myra Jehlen, Routledge, 1997); and *Fear of a Queer Planet: Queer Politics and Social Theory* (University of Minnesota Press, 1993).

Jedediah Britton-Purdy is the Raphael Lemkin Professor at Duke Law School. He was previously the William S. Beinecke Professor at Columbia Law School. He is the author of seven books, most recently *Two Cheers for Politics: Why Democracy Is Flawed, Frightening, and Our*

Best Hope (Basic Books, 2022). Other titles include *This Land Is Our Land: The Struggle for a New Commonwealth* (Princeton University Press, 2019) and *After Nature: A Politics for the Anthropocene* (Harvard University Press, 2015). He has contributed to publications such as *The Atlantic*, *The New Republic*, *The New Yorker*, *The Nation*, *Democracy Journal*, *Dissent*, and *Jacobin*, and also recently edited the Norton Library edition of Thoreau's *Walden and Other Writings* (2023).

Dale Jamieson is Professor Emeritus of Environmental Studies at New York University, where he is also Affiliated Professor of Law, Medical Ethics, and Bioethics, a founder and former chair of the Environmental Studies Department, Founding Director of the Center for Environmental and Animal Protection, and former Professor of Philosophy. He is the author of *Reason in a Dark Time: Why the Struggle to Stop Climate Change Failed—and What It Means for Our Future* (Oxford University Press, 2014), *Ethics and the Environment: An Introduction* (Cambridge University Press, 2008; second edition in press), *Morality's Progress: Essays on Humans, Other Animals, and the Rest of Nature* (Oxford University Press, 2002), and most recently, *Discerning Experts: The Practices of Scientific Assessment for Environmental Policy* (University of Chicago Press, 2019), coauthored with Michael Oppenheimer, Naomi Oreskes, and others. He is also the coauthor of *Love in the Anthropocene* (OR Books, 2015), a collection of short stories and essays written with the novelist Bonnie Nadzam.

Michael Lucey is the Sidney and Margaret Ancker Distinguished Professor of Comparative Literature and French at the University of California, Berkeley. His most recent books are *What Proust Heard: Novels and the Ethnography of Talk* (University of Chicago Press, 2022) and *Someone: The Pragmatics of Misfit Sexualities, from Colette to Hervé Guibert* (University of Chicago Press, 2019). He has edited a special issue of *Paragraph* on "Approaching Proust in 2022" and coedited a special issue of *Representations* on "Language-in-Use and the Literary Artifact." His most recent translation is of Didier Eribon's *The Life, Old Age, and Death of a Working-Class Woman* (Semiotext(e) and Penguin, 2025), and a new translation of André Gide's *The Counterfeiters* will be published by the University of Chicago Press.

Anahid Nersessian is Professor of English at UCLA and the poetry editor of *Granta Magazine*. She is the author of three books, *Utopia, Limited: Romanticism and Adjustment* (Harvard University Press, 2015), *The Calamity Form: On Poetry and Social Life* (University of Chicago Press, 2020) and *Keats's Odes: A Lover's Discourse* (University of Chicago Press, 2021; Verso, 2022). A frequent contributor to *The New York Review of Books*, she has also published essays and reviews in *Critical Inquiry, ELH, NLH, The London Review of Books, New Left Review, Bidoun, Mousse Magazine, The Times Literary Supplement*, and elsewhere.

Introduction

MICHAEL LUCEY

Every so often a heavy truck or a bus careens down the street we live on, traveling at just the right speed and on just the right trajectory that it hits an unevenness in the road surface and gives the two-story building where our apartment is located a particular kind of jolt. That jolt triggers the seismic shutoff valve for our gas line. Typically, we aren't immediately aware that this has happened. There are a number of ways we might learn of it, usually highly annoying and inconvenient ones. Perhaps we wake up a bit late and are trying to get in a quick shower before heading off to work, a shower that starts out tepid and moves rapidly toward cold. Or perhaps it's the end of a long day and we want to rustle up something quick for dinner, only to discover the burners won't light, requiring a trip down to the street (more often than not when it's damp outside or raining) with a flashlight and a small screwdriver, to lift the cement cover up off the box underneath the sidewalk, reset the valve, then dash back upstairs to light the stove so the gas flows, and then down to the basement crawl space to relight the pilot on the water heater. We also have a gas furnace up in the attic, but it has its own self-lighting pilot, so thankfully it takes care of itself. And given that it is always a bit chilly in San

On the Grid. Michael Warner, Oxford University Press. © Regents of the University of California 2025.
DOI: 10.1093/oso/9780197696248.003.0001

Francisco (and in our apartment), it's not usually by way of the furnace not functioning that we notice our mishap.

The Tanner Lectures on Human Values that Michael Warner delivered in Berkeley in March 2018 focus on the history of the generation of electricity and its distribution via the grid, on what access to power from the grid does to us as ethical agents in relation to the environment, and on how the goal of decarbonizing the electricity grid fits—or rather fails to fit—with long-standing environmentalist values. Warner notes in his first lecture that gas distribution preceded the distribution of electricity and served as a model for it. Gas distribution had water as a prior example: "Before electric wires and gas jets came inside the home, water and sewer engineers and corporations had begun carrying resources into and waste out of cities by means of pipes, pumping stations, aqueducts, faucets, and drains. The unconscious end user, defined by continuous potential of demand and distance from resources, was thus built by successive layers into the techno-social environment" (p. 42–43). Grids have as an aim not to be noticed, just to be available for use. Something has gone wrong when you notice them.

As Warner puts it, grids—electric, gas, water—have "the ability to disappear" (p. 15). Normally, we just use what they offer us without giving it much thought. They are, under usual circumstances, "engine[s] of unconsciousness" (p. 30), and, as Warner memorably puts it, "We actively dislike being called out of this empowering sleep" (p. 32). His example of dislikeable moments are ones in which a substation or a wind farm forces itself on our attention: "We feel entitled to our obliviousness" (p. 32). A shower unexpectedly fading to cold or a stove burner that won't light are similar affronts to our obliviousness.

Now if we have a seismic shutoff valve for our gas line in the first place, it is because, in the seismically active region in which I live, we do have reason to think, at least from time to time, about certain dangers associated with being connected to grids—the gas grid in particular. (And not all the dangers associated with this grid, as it turns out, are related to seismic activity.)[1]

In the fall of 2018, Northern California experienced a devastating wildfire season, including, some eight months after Warner delivered the lectures reprinted here, the deadly Camp Fire, caused by a "poorly maintained electric transmission line."[2] Air quality issues related to the wildfires caused the Berkeley campus to shut down for several days that fall. There would be wildfire-related air quality issues in fall 2019 (and again in fall 2020, although that was during a semester of remote instruction), but the campus was actually closed for several days in 2019, not because of an actual wildfire, but because the beleaguered utility, PG&E, had instituted a policy of emergency power outages to mitigate the risk of fire outbreaks during especially dangerous moments of the wildfire season; the campus was thus cut off from the electricity grid for several days on a couple occasions that fall.

A cold shower, a stove that won't light, a furnace that isn't doing its job, the prospect of regular class cancellations during obligatory power outages in the years ahead, a massively destructive fire with significant loss of life: There are ways these days in which grids remind us that they are there. And yet. One of the core issues Warner and his respondents all take up is the fact that the grid-enabled capacity to use natural resources without thinking much about doing so in our current era of climate change results in "a conceptual impasse about the kind of agency involved in anthropogenic climate change" (p. 14), a question of how to act under present circumstances. Does anything I do really make a difference? As Warner insists, "To think about climate change requires us somehow to bring the grid and our

1. See, for example, https://en.wikipedia.org/wiki/San_Bruno_pipeline_explosion.
2. https://en.wikipedia.org/wiki/Camp_Fire_(2018).

dependence on it into consciousness, into shared concern, and into politics" (p. 29), but as Dale Jamieson adds somewhat bluntly, many of us mostly experience "episodes of guilt against a broad background of a sense of complicity that is not systematically connected to action. And so together we destroy the world but no one feels responsible" (p. 84).

Where do meaningful action and responsible behavior lie? Authorities in the San Francisco Bay Area plan to phase out the use of gas furnaces and gas water heaters over the next several decades.[3] When I wonder uncomfortably as I reset my seismic shutoff valve about what I might do about my gas furnace, my gas water heater, and my gas stove, I find myself very clearly within the conceptual space that Warner's lectures and the three responses to it are exploring. The city and the region where I live want me to electrify everything (or at the very least, the furnace and the water heater) as part of a project of reducing the emissions gas appliances produce. Now Warner demonstrates in his first lecture that "as a strategy for both decarbonizing and expanding the reach of the power grid, *electrify everything* overturns most of the received wisdom of environmentalism" (p. 21). It has no necessary connection to sustainability (obviously not all electricity is currently produced in an emission-free manner from renewable resources) or to conservation. Still it is true that I pay a little extra each month through CleanPowerSF's SuperGreen service so that I can (in a rather metaphorical manner of speaking) "power [my] home with 100% renewable electricity."[4] And I could decide, when it is time for us to replace the water heaters in our two-unit building (and when, most likely, gas water heaters will no longer be an option) also to add some solar panels to our

3. See https://www.sfchronicle.com/climate/article/bay-area-regulators-vote-end-sales-key-gas-home-17836615.php and https://www.sfenvironment.org/electrify.
4. See https://www.cleanpowersf.org/supergreen.

roof and some battery storage for the building. PG&E now encourages this kind of thing on its website.[5] I have friends who quite enjoy tracking on an app how much energy their solar panels produce and send back to the utility to lower their electricity bills. As Warner writes, "On the grid as it is currently designed, the meter is the only means by which you are made aware of your power. If you look at it at all, your actions are translated back to you as usage alone" (p. 41). The point is well taken, even if these days some people enjoy watching their meter run, so to speak, backward now and then. Surely there are environmental ethics operative to a degree in the desire of many homeowners around California to install solar panels and lower their PG&E bills, but perhaps the virtues of conservation, of pollution reduction, and of environmental justice in general do seem somewhat obscured by economic interests, as well as by a desire for a kind of hybrid energy independence that doesn't imply being fully cut off from the grid.

"When you buy a house," Dale Jamieson notes in his response to Warner, "it comes with ceilings, floors, walls, fixtures, and probably appliances. It could also come with an energy system" (p. 83). Given all the retrofitting that would be required in our building to get rid of the gas heating systems, the gas water heaters, and maybe even the gas stoves and ovens (we know that cooking with gas has become a big ideological flashpoint of late—the city of Berkeley has recently been made an example of in this regard),[6] why would I not see if we could install enough in the way of solar panels and batteries to have something approaching our own self-sufficient energy

5. See https://www.pge.com/en/clean-energy.html. They do recommend, since "solar panels can last more than 25 years," that you "make sure your roof is no more than seven years old to avoid reinstalling panels a second time" (https://www.pge.com/en/clean-energy/solar/getting-started-with-solar.html). Our roof is more like twenty-five years old, so we would probably want to think about timing a new roof with the installation of panels.

6. See https://www.berkeleyside.org/2024/01/03/berkeley-gas-stove-ban-ruling.

system? PG&E itself links home battery storage options to the need some consumers might have to keep medical equipment or refrigeration or air conditioning running during an emergency outage such as may happen from time to time in high-risk fire areas. (Certainly, a lot more people and businesses around here have acquired generators over the past few years—mostly, I imagine, propane- or gasoline-powered ones—given the increased likelihood of outages of various kinds.)

Suppose, in our best attempt at pursuing some version of the "green virtues" of which Jamieson speaks (see p. 93), we did electrify everything, install some solar panels (along with a new roof), make everything as efficient as possible—it does then seem, based on my recent perusal of a few solar websites, like my downstairs neighbor, my husband, and I could probably have a more or less energy-independent building. (We might have to pay close attention to our usage in December and January.) Or would we? As Warner cautions his reader:

> Do you know what fuel powers the room you are in right now? The odds are very high that you are burning fossil fuels by reading this, but no one needs to make the connection, and when it comes to internet use, hardly anyone ever does. (p. 23)

Maybe I would have solved the problem of the electricity for the lights and for whatever screen I was reading on, if I wasn't reading a hard copy. But what if I was streaming music as I read, or what if I stopped to look up one of the article referenced in a footnote, or did some internet searching on this topic or that, or somehow, whether I knew it or not, whether I wanted to or not, made use of some AI-enhanced feature of my telephone or tablet or laptop—features that will have been integrated into many of those devices by the time you read this volume.

I'm sure many people reading this will have bought carbon offsets for air travel you have done. We will all also have probably read the criticisms of that entire industry.[7] We might wonder about the idea of offsets for any music or video streaming that we do these days. The debate over the relative sizes of the carbon footprints of music on vinyl or on CD or streaming (or live for that matter) is itself confusing.[8] Can you trust that Spotify's switching to the Google Cloud Server has moved it toward carbon neutrality? Is it better to download tracks you listen to repeatedly instead of streaming them over and over? Just for the record, I listened to music on CDs while working on this introduction, but I have no real confidence that it was virtuous to have done so. I do, also, stream music regularly and find it amazing that I can, should I wish, have so many different versions of George Shearing's song "Conception" (multiple Shearing versions, Bud Powell's, Miles Davis's, Art Blakey and the Messengers', Bill Evans's, Keith Jarrett's) at my fingertips in the space of a single short listening session with so little effort. It is when I am streaming music in this way that I feel most like the "user" of the grid that Warner is calling our attention to: "The constancy of the grid is, for users, the constancy of their own ability to do things that involve electric

7. This was a hot topic in the *New York Times* in May 2024. See, for example, Manuela Andreoni, "Can Forests Be More Profitable Than Beef?," May 2, 2024, https://www.nytimes.com/2024/05/02/climate/amazon-reforestation.html; Susan Shain, "Are Flight Offsets Worth It," May 6, 2024, https://www.nytimes.com/2024/05/06/climate/nyt-questions-flight-offsets.html; Brad Plumer, "Carbon Offsets, a Much-Criticized Climate Tool, Get Federal Guidelines," May 28, 2024, https://www.nytimes.com/2024/05/28/climate/yellen-carbon-offset-market.html; and Manuela Andreoni, "How to Fix Carbon Offsets," May 28, 2024, https://www.nytimes.com/2024/05/28/climate/how-to-fix-carbon-offsets.html.

8. Cf. Amanda Petrusich, "The Day the Music Became Carbon-Neutral," *New Yorker*, February 18, 2020, https://www.newyorker.com/culture/culture-desk/the-day-the-music-became-carbon-neutral; "The Political Ecology of Spotify," September 28, 2020, https://mastersofmedia.hum.uva.nl/blog/2020/09/28/the-political-ecology-of-spotify-how-streaming-takes-a-toll-on-the-environment/; "The Environmental Impact of Music Streaming, Explained," https://www.cbc.ca/music/the-environmental-impact-of-music-streaming-explained-1.6843948.

power, at will and whenever they like" (p. 32); "*the user*, a private subject of maximal demand and minimal awareness" (p. 41). That's me, listening contentedly to version after version of "Conception."

On the other hand, I have personally, and probably for reasons that are utterly predictable from a sociological point of view, yet to directly query to ChatGPT on any topic. (I have, I should admit, used Google Translate on a significant number of occasions, since my German has gotten quite rusty, and I also recently needed help with a few documents in Croatian.) I saw on social media and recently found a citation in a *Scientific American* article the catchy factoid that "every 10 to 50 responses from ChatGPT running GPT-3 evaporate the equivalent of a bottle of water to cool the AI's servers."[9] While I may currently be attempting to be virtuous by avoiding ChatGPT and its ilk, it is becoming more and more difficult to do so. AI is more or less unavoidable if you search the internet with almost any search engine, and it will soon be next to unavoidable on, for instance, Apple devices with built-in Apple Intelligence.[10] We increasingly (when we use the internet, and especially AI, as when we stream videos or music) consume electricity and water that we are mostly unaware of.

We could just decide to trust that Apple, Google, Microsoft, Meta, Amazon, and so on will themselves achieve meaningful versions of carbon neutrality in the near future, and so, so to speak, go back to sleep and be happy users who trust others will have solved

9. Allison Parshall, "What Do Google's AI Answers Cost the Environment?," June 11, 2024, https://www.scientificamerican.com/article/what-do-googles-ai-answers-cost-the-environment/. The study Parshall cites is Pengfei Li et al., "Making AI Less 'Thirsty': Uncovering and Addressing the Secret Water Footprint of AI Models," October 29, 2023, https://arxiv.org/pdf/2304.03271. See also Joyeeta Gupta et al, "AI's Excessive Water Consumption Threatens to Drown Out Its Environmental Contributions," *The Conversation*, March 21, 2024, https://theconversation.com/ais-excessive-water-consumption-threatens-to-drown-out-its-environmental-contributions-225854.

10. Tripp Mickle, "Apple Jumps into A.I. Fray with Apple Intelligence," *New York Times*, June 10, 2024, https://www.nytimes.com/2024/06/10/technology/apple-intelligence-openai.html.

"the problem." Apple's "2024 Environmental Progress Report" sets the following goals, for example:

> Achieve carbon neutrality for our entire carbon footprint by 2030—reducing related emissions by 75 percent compared with 2015.
>
> Transition our entire value chain, including manufacturing and product use, to 100 percent clean electricity by 2030.
>
> By the end of fiscal year 2025, we plan to have certified all Apple-owned data centers in the Alliance for Water Stewardship Standard.[11]

(But note one recent article's observation that "Microsoft pledged four years ago to bring its greenhouse-gas emissions to zero [or even lower] by the end of the decade. But the company's recent sustainability report shows that instead, emissions are still ticking up, and some executives point to AI as a reason.")[12] Or we could accept that the concerns about "ethicalized awareness" (p. 76) that Warner lays out in the following lectures are more compelling and pressing than ever, that we still face the crucial and difficult task of "untangling means and ends in modern infrastructure" (p. 58).

My goal in this introduction has been mostly to foreground a state of perplexity I imagine many of us live in regarding a contemporary experience of gridded infrastructure, and to set a mood, probably a slightly anxious one, in which Warner's lectures and the three compelling responses by Dale Jamieson, Jedediah Britton-Purdy, and

11. "2024 Environmental Progress Report," 5, 112 n. 15. Available at https://www.apple.com/environment/pdf/Apple_Environmental_Progress_Report_2024.pdf.

12. Casey Crownhart, "AI Is an Energy Hog. This Is What It Means for Climate Change," *MIT Technology Review*, May 23, 2024, https://www.technologyreview.com/2024/05/23/1092777/ai-is-an-energy-hog-this-is-what-it-means-for-climate-change/.

Anahid Nersessian can take on their full sense of urgency, asking us to confront (and to improve) the extent of, in Nersessian's words, "our epistemic grasp on griddedness" (p. 124).

Interestingly, all three of the responses to Warner's lecture sound similar notes. Notice Jamieson's assertion that "a picture that views individual people as the primary protagonists—in their various roles and relationships, with their own sense of the good and sources of meaning—is one that is more natural to the climate change story than one that sees it as a problem in search of a solution from economics" (p. 93). This is a corollary to his view that "developing, inculcating, and acting on such green virtues as humility, mindfulness, cooperativeness, and respect for nature are important steps along the way" to "the recovery (or reconception) of agency" (p. 93). Britton-Purdy sets out to "inquire into the prospect of *generating* any kind of collective agency, any kind of *we*, in the face of an infrastructure world that relentlessly ties people together" (p. 98). This requires of people a "vividness of vision" that would contribute to the viability of "a political community . . . in which it remains possible, as Thoreau put it in 'Civil Disobedience,' to appeal against the people to themselves—to make the claim that we could do, and be, something else, that we carry in ourselves the possibility of a different world" (p. 105). Finally, Nersessian turns to poets and artists who might help us to "enlarge our sense of what a differently gridded future might make possible"(p. 125). At one point, drawing on the writings of Lakota historian Nick Estes, she asks what kinds of political awareness might "make possible a new social, economic, and ecological model in which our relationships to one another and to the land are not occluded by technology but used to negotiate it, and in which the collective might supplant the individual as both subject and object of ethical action" (p. 115). Imagine that.

ON THE GRID: CLIMATE CHANGE AND THE UTOPIA OF GREEN ENERGY

Preface

The most shocking thing about climate change—more shocking, in a way, than the apocalyptic scenarios of popular imagination—is the way life just goes on. As the planet burns, one keeps putting gas in the tank, flicking on the lights, checking email, watching TV, running the AC, eating factory-farmed food, ordering deliveries from Amazon, charging the phone, and dreaming of that next trip to the beach. Everything happens as though no one notices. As David Wallace-Wells notes, more than half of all the carbon dioxide that humans have put into the atmosphere has been put there in the last three decades, which means that humans have done more to alter the planet's climate in the years since Al Gore's *Earth in the Balance* than in all of human history before.[1] How is this possible?

These lectures grew out of this apparent disconnect between awareness and action. "Everybody talks about the weather, but nobody seems to do anything about it," as Charles Dudley Warner liked to say. (The joke is often credited to his friend Mark Twain, probably wrongly; every good quip gets attributed to Twain.) What he noticed about weather is equally true of climate. Knowledge of planetary crisis has done little to change the habits of everyday life

1. David Wallace-Wells, *The Uninhabitable Earth: Life After Warming* (New York: Tim Duggan Books, 2019), 4.

On the Grid. Michael Warner, Oxford University Press. © Regents of the University of California 2025.
DOI: 10.1093/oso/9780197696248.003.0002

that brought the crisis about. Actions that do lie within the reach of ordinary people, like recycling, feel like placebos. What is to be done?

Dipesh Chakrabarty, in his own Tanner Lectures, observes that this disconnect has much to do with a conceptual impasse about the kind of agency involved in anthropogenic climate change. Where we like to think of action as conscious, intentional, and personal or collective, the species- or system-level agency behind greenhouse gases is mostly unconscious, unintended, and distributed.[2] In these lectures I try to bring into focus this general disconnect by examining a specific case of action: the current ambition of decarbonizing modern power systems, primarily through the large-scale project of replacing fossil fuels with renewable electricity. More and more of the power we use, the thinking goes, needs to be driven by an electrical grid that in turn needs to be fueled more and more by wind, solar, and other greenhouse-neutral sources. Get rid of internal combustion cars and trucks, for example; replace them with electric ones and charge them with power from wind and solar and other nonfossil sources. In this projected transformation lies the greatest hope that humanity—especially that part of humanity living in advanced industrial societies—might actually do something to mitigate the harm that results from modern ways of life. The change called for by this strategy is imagined as collective, conscious, intentional action. But power use itself—driving around in cars, to stick with this example—is distributed, mostly unconscious and unintentional, while the fueling of electrical generation itself seems opaque and distant. You think about going down to the store, not about using energy. Where the energy comes from, were you to think about it, would be mysterious and remote. And thus you keep doing it because it hardly seems possible to do otherwise. What might be thought of as agency is so baked

2. Dipesh Chakrabarty, "The Human Condition in the Anthropocene," in *The Tanner Lectures on Human Values*, vol. 35, ed. Mark Matheson (Salt Lake City: University of Utah Press, 2016).

into modern infrastructure that it can only be brought into awareness or reflective action by extraordinary effort, in fleeting moments of apparently futile recognition. People feel powerless to do much about the power they use—where it comes from, what they do with it, or what harm it causes. And so life goes on, pretty much as before.

The disconnect, I will suggest, has much to do with the way power is mediated. In the ordinary case one has no awareness of how power is generated or becomes available for use. It is the background of actions that seem to be about something else, like staying warm or listening to music. And despite the vast and heroic scale of the green power transition that represents humanity's best hope, the greening of the grid is unlikely to do much to change the essentially distributed and unconscious nature of environmental action.

The reason is the grid itself. Electrical infrastructure has become a pervasive and defining feature of modern life. To be deprived of it is to be excluded from modern rights, freedoms, and forms of sociality. Yet it is a remarkable absence in both ordinary consciousness and sophisticated critical thought about modern culture and politics. The vast corpus of critical social theory has mostly failed to notice it, just as you probably don't notice the power you are using while reading these words. The grid, like all infrastructure, has the ability to disappear. It is designed that way. More: The electrical grid, for reasons I will sketch in the first lecture, is an engine of ordinary unconsciousness. It is set up to discourage anyone from being aware of it. Still more: To the degree that it becomes manifest in ordinary life, it is by enabling comfort, convenience, and capability. Thus we tend not to notice our own attachment to it, or even our feeling of entitlement about not noticing.

It has not always been so. Before climate change came to attention as even a possibility, let alone a reality, people were conducting a wide range of experiments in alternative energy. These were mostly small-scale, fringe endeavors, and that was part of their point. In the second half of the twentieth century, solar architecture, photovoltaic electricity, wind power, and other infrastructural innovations like septic tanks were developed for domestic use mainly by people who thought of them as ways of being more conscious and more in control of the environmental conditions of sustainable ways of living. This meant bringing them down in scale, avoiding the dependence on corporate utility structures, facilitating proximity between power generation and end uses. From midcentury until the 1980s, this culturally and politically diverse set of experiments came to be known as "offgrid."

That project now seems archaic and quaint, when it is remembered at all. In the second lecture I review this period of infrastructural experiment, with a couple of aims in view. The first is simply to illuminate by contrast what grid mediation entails; you can see the grid better if you imagine not being on it. The second is to underscore what I think is a mostly unnoticed transformation in the environmental imagination. Those offgrid experiments carried with them hopes and understandings that were virtually synonymous with twentieth-century environmentalism: raising awareness of one's relation to the earth; taking responsibility for one's resources and waste; organizing collectively to interrupt the absorption of life by corporate capitalism; questioning the default mandates of consumption, comfort, and convenience; living with attunement to the elements; practicing stewardship. None of these aspirations have much traction on the green energy transition any longer, and it is possible to see historically where environmental understanding and infrastructure parted company: in the very years when climate change began to be a concern, in the 1980s and 1990s. This is ironic, to say the least, and it is an irony that tends not to be noticed when we think of climate change concern as continuous with longer histories of environmentalism.

By revisiting the offgrid episode in the second lecture I also mean to be doing some auto-critique. My own formation by environmentalism came about in the context of the offgrid movement. I vividly remember the excitement occasioned by the *Whole Earth Catalog* in its several iterations. And while the alternative networks around that publication themselves transformed into something entirely different—ultimately becoming the germ of Silicon Valley, as Fred Turner has demonstrated—the offgrid commune movement had (and has) ongoing significance for me because of my involvement with Radical Faeries.[3] For those of us who stood in the legacy of Harry Hay and his collaborators, queerness and the sociotechnical experiments of environmental awareness were linked, both in practice and in aspiration. The Faerie sanctuary that is one of my primary communities was founded in the early 1970s as an offgrid alternative commune and remains so, though the frontier of separation between it and the surrounding culture of consumption has been progressively softened as newer generations bring their smartphones and computers to the land. In these lectures I mean to be clearing space and taking stock, and I might as well admit that the contingencies of my personal history have a good deal to do with the way the reflection takes shape here.

For this publication I have tried to stick as closely as possible to the shape of the lectures as they were delivered at Berkeley, though I have tried to clarify and expand where it seemed necessary, especially in the second lecture. Because they were intended as lectures, I have tried to keep footnotes to a minimum, though I have added some for the benefit of anyone who would like to read further in the wide range of topics I cover so hurriedly.

3. Fred Turner, *From Counterculture to Cyberculture: Stewart Brand, the Whole Earth Network, and the Rise of Digital Utopianism* (Chicago: University of Chicago Press, 2016).

Lecture I

On the Grid

MICHAEL WARNER

Climate change is so daunting, and in many ways so vindicates the long tradition of environmental concern, that it is easy to overlook how the problem of greenhouse gas emissions has changed the scheme and direction of environmental concern itself. In these lectures I will explore the example of the electric power grid as an object of concern. Given that many people understand climate change as entailing an obligation to do something about carbon emissions, what difference does it make that most of our carbon emissions are mediated by the grid? In the second lecture I will take up the history of debate about environmental ethics, which has a long tradition of asking what our moral obligations of environmental care are and why we should care. That question has generally been posed in a framework of ethical philosophy that brackets infrastructure and its regulation; the focus has been on individual moral agents. What must I do, and why? I will ask how the posing of this question changes when we consider the infrastructural unconscious. To look ahead briefly, I will be asking how infrastructure helps to constitute us as ethical subjects in the first place. To do so is to invoke "ethics" in something more like its Foucauldian sense—asking not just what a moral code requires of

On the Grid. Michael Warner, Oxford University Press. © Regents of the University of California 2025.
DOI: 10.1093/oso/9780197696248.003.0003

individuals, but how individuals can in different ways be related (or not) to spheres of moral action—than the questions posed by the prevailing traditions of Anglo-American normative ethics. In those traditions, the mediation of action by large-scale infrastructure has mainly surfaced as a reason for skepticism. Climate change is beyond our control; it requires governmental action, while individual choices seem too marginal and inconsequential to do much; and the various end states proposed by environmental ethics—passing beyond human chauvinism in order to arrive at sustainability, for example—can dispose us to think that the framework of individual moral choice is part of the problem rather than the solution. By starting with an analysis of what the grid *is*, I intend to deepen such concerns rather than to dispose of them.

The question of the grid has been reframed by the rise of renewable sources of energy. Until quite recently, solar and wind power were deployed mainly for offgrid sites, in both the developed and undeveloped worlds; the grid was almost entirely tailored to fossil fuel interests, along with nuclear and large-scale hydropower, and for the most part it still is. Now, however, in the age of climate change, it is the grid itself that is beginning to be rebuilt and reregulated for wind, solar, and other forms of renewable energy. Where environmental concern in the twentieth century led away from utilities and fossil fuels, to the point that "offgrid" came to connote an ethically supercharged way of being for those who could elect it, that same concern now leads to hope for a bigger, less centralized, better integrated, more flexible grid—a grid that will be, above all, *smart.*[1]

In short, the grid is good, and the stakes of this new utopia could hardly be higher. While climate change mostly overtakes us as weather, a change in our environment over which we have no

1. The best account of this hope is Peter Fox-Penner, *Smart Power: Climate Change, the Smart Grid, and the Future of Electric Utilities* (Washington, DC: Island Press, 2010; rev. ed. 2014).

control, decarbonizing the power grid has become the main *practical* context for responding to the crisis of greenhouse gas levels. The grid is both local (we are surrounded by its wires right now) and the closest thing we have to an opportunity for planetary action. If carried through, the greening of the grid would represent the largest achievement of the environmentalist movement—too little and too late to save the earth as we knew it in the Holocene, certainly, but nevertheless the main hope for social adaptation to the Anthropocene. Notwithstanding the opposition of climate denialists (who thrive, interestingly, almost exclusively in the Anglosphere), commitments are now being made—shaped in part by the Paris Accord of 2016 and involving a global brokering of "stakeholder" interests—that in effect double down on the electrical power grid as the future of humanity in the Anthropocene.

And it is not hard to see the reason for this focus. The electric power sector is currently the largest contributor to CO_2 emissions; close all the coal-powered power plants and you have solved the most glaring part of the escalating problem. And the importance of the power grid far exceeds its present share in emissions: Efforts to reduce emissions in transportation, the second-highest sector, hinge on the expected shift to electric vehicles—a shift that will only be "green" to the extent that the power needed to charge them comes from renewable sources. (Electric vehicles are essentially part of the grid since their batteries store power in off-peak hours for use in peak hours, thus potentially providing critical load management.) The most plausible strategy for cutting greenhouse gas emissions, therefore, is the one that has been summed up with disarming simplicity by the journalist David Roberts: "Electrify everything."[2]

2. http://www.vox.com/2016/9/19/12938086/electrify-everything. See also Keith Dennis et al., "Environmentally Beneficial Electrification: The Dawn of 'Emissions Efficiency,'" *Electricity Journal* 29, no. 6 (July 2016): 52–58; Dennis and coauthors argue that emissions

As a strategy for both decarbonizing and expanding the reach of the power grid, *electrify everything* overturns most of the received wisdom of environmentalism. Twentieth-century environmentalism sought to raise popular consciousness, mobilize for collective political action, promote transparency, foster ethical concern for sustainability, end blind consumerism through the cultivation of ascetic virtues (conservation, stewardship, thrift), and build change through the aggregated voluntary efforts of individuals. *Electrify everything* needs none of those commitments. That is part of its strength, since it is a strategy that does not depend on widespread ethical reformation or even awareness and does not call on most people to give anything up. By comparison, when Bill McKibben published *The End of Nature* in 1989, helping to put global warming at the center of attention, the solution he proposed was "a more humble way of living." Barely noticed at the time were subtle but significant changes in infrastructure and its regulation that now, thirty years later, have arguably far greater consequence than such a program of evangelical humility, or indeed than any of the subjective reorientations that activists and academics like to think of themselves as engendering.

In contrast to more familiar environmental causes like recycling, which succeed or fail depending on the extent of popular participation, an infrastructural politics of decarbonization raises questions about the opacity of technical change to politics. It requires policy-setting, certainly, but the direction of change tends to be set in places that are not at all planetary, and have all the heroism and glamour of a utility board meeting.[3] At the same time that many of us worry about the power of technocratic elites, we tend to assume

efficiency requires electrification even if it results in lower energy efficiency, a distinction not often made in policy discourse.

3. As Gretchen Bakke observes, most of the regulatory structures give priority to stakeholders with profit motives. Bakke, *The Grid: The Fraying Wires Between Americans and Our Energy Futures* (New York: Bloomsbury, 2017), xxv.

that somewhere the engineers are working on solutions—now more than ever. Moreover, all parties seem to agree that it is the price drop in renewables that contributes momentum to the shift, willy-nilly. Until very recently most commentators dismissed as fanciful speculation the idea that renewable energy would soon play a significant role in mitigating climate change. But the cheapness of solar and wind changes the equation, as the saying goes. In the United States, solar and wind accounted for more than two-thirds of all new electric capacity in 2015.[4] In 2017, the solar market was ten times as large as it was in 2010.[5] As they get cheaper and more efficient, solar power and wind (almost all of the latter in industrial-scale wind farms) spread faster, and by economies of scale become cheaper still.[6] This trend has a political element, since it is driven in part by tax credits and international climate policy—that is, the opaque politics of grid regulation and marketized energy. Indeed, precisely because it is so opaque and indirect, the processes of grid reform may be buffered to some degree from the climate-denialist politics and obstruction (or at least, such is the hope of many observers) to the degree that the shift to renewables seems unpolitical, driven by economic and technical imperatives.[7] Decarbonizing the grid, in short, is being envisioned in ways that mostly bypass models of popular mobilization, whether by necessity or choice. It is coming to look more and more

4. Joe Ryan, "A Renewables Revolution Is Toppling the Dominance of Fossil Fuels in U.S. Power," Bloomberg News, February 4, 2016. http://bloom.bg/1P8eSpi.
5. Jonathan Chait, "Climate Change and Conservative Brain Death," *New York Magazine*, March 14, 2016, http://nymag.com/daily/intelligencer/2016/03/climate-change-and-conservative-brain-death.html.
6. Chris Mooney, "Solar Power Is Poised for an Unforgettable Year," *Washington Post*, March 2, 2016, https://www.washingtonpost.com/news/energy-environment/wp/2016/03/02/solar-energy-is-poised-for-an-unforgettable-year.
7. See for example the Morgan Stanley report "The US Election: Impacts to Clean Tech and Utilities Skew Positive," July 27, 2016, summarized at https://web.archive.org/web/20161212111357/https://www.morganstanley.com/ideas/clean-energy-trump.

like weather. *Electrify everything* relies on the grid to mediate climate action, just as it aims to have the grid—smarter, decentralized, more efficient—mediate everything else.

The layer of arbitrariness between you and your fuel is, ironically, the ground of hope for reducing carbon emissions. The grid is designed so that you never have to know what the fuel source of your energy is. It combines oil, coal, gas, hydro, nuclear, biomass, geothermal, wind, and solar sources into a single, undifferentiated medium for unspecified capabilities. That is both what makes it possible to shift the sources and what makes it possible not to know or care about the fuel source—and thus carbon emissions—when you use electricity. Do you know what fuel powers the room you are in right now? The odds are very high that you are burning fossil fuels by reading this, but no one needs to make the connection, and when it comes to internet use, hardly anyone does. The grid makes this arbitrary relation to resources a practical opportunity: As more and more power is generated from renewables, it gets easier to reduce carbon emissions without reducing energy consumption. Denmark, for example, has a per capita energy consumption comparable to other European nations, but it leads the world in the use of wind power, making carbon emissions much lower per capita. Conversely, China's per capita energy consumption is still rather low because of its large and mostly poor population, but reliance on coal in the Chinese grid means that its total grid emissions are even worse than those of the United States. Whether your power is green or not has very little to do with how much of it you use.

Some countries, of course, have much more grid-dependent ways of life than others. The United States outranks almost everyone else by this standard, which (along with a landscape built out for automobiles) is a major reason why US per capita CO_2 emissions stand among the highest in the world. To many, this consideration reinforces the sense that energy use is an ethical issue: Americans, it seems, are pigs. As long as that level of consumption requires fossil fuels, concern about greenhouse gases should require using less. But what if it doesn't? The goal of 100 percent renewable energy is nowhere close to being realizable, but as a goal it no doubt helps to neutralize concern about energy consumption itself in the long run. Climate change does not result directly from an energy-intensive society; it does so only because fossil fuels release CO_2 when burned. It so happens that most electric generation currently consumes fossil fuels; if it did not, then high consumption would not be an urgent issue of climate change, regardless of what other reasons one might have to reduce it through virtue.[8] Ethical or political arguments for reducing energy, in other words, lose their purchase on climate change to the extent that energy generation can be shifted away from carbon-intensive fuels; the problem is the fuel source, not the end use. This is why the major source of optimism about climate change, to the degree that anyone has any, lies not in creating a less demanding public but in retooling the grid.

The engineers, activists, think tanks, regulators, and industries leading the current transnational energy shift, citing environmental urgency and technical breakthroughs, are not waiting for ethical change. Amory Lovins, founder of the Rocky Mountain Institute and for four decades the leading guru of energy reform, writes encouragingly that "we humans have begun the most important infrastructure shift in history, melding energy with information technology and design, blending technical with social breakthroughs, creating one astonishment after another."[9] He also states baldly that his plan "assumes no lifestyle changes that could materially reduce comfort

8. For this and other reasons having to do with scale, Dipesh Chakrabarty argues that the planetary issues of climate change are not well captured by the analytics of capitalism. "Climate and Capital: On Conjoined Histories," *Critical Inquiry* 41 (Autumn 2014): 1–23.

9. Amory Lovins, *Reinventing Fire: Bold Business Solutions for the New Energy Era* (New York: Chelsea Green Publishing, 2013), xiii.

or convenience."[10] Grid reformers like Lovins have learned in this way to sidestep the very idea that consumption should be ethically regulated. To them such appeals, characteristic of twentieth-century environmental thought, translate as a minor form of "demand reduction." Far more important than voluntary, conscious conservation, they argue, is efficiency. Better to install efficient light bulbs than train consumers to turn them off.[11]

In arguments for environmental justice, it is very common to encounter critiques of consumption in the developed world; nevertheless, justice concerns also commonly lead for different reasons to the same strategy: *Electrify everything*. Over a billion people on earth still have no electricity. Since 2000 the UN has been targeting energy services as critical to eliminating poverty; both human rights and climate justice arguments have been made for universal energy access.[12] In the UN's "Sustainable Development" guidelines, this is goal 7: "Ensure access to affordable, reliable, sustainable, and modern energy for all."[13] A number of left scholars have emphasized the unequal politics of grid construction; in places like South Africa and Israel/Palestine, for example, the grid was a colonial project that helped to build racial inequalities.[14] Many activists advocate novel

10. Lovins, *Reinventing Fire*, x. Lovins himself, interestingly, lives in a house designed to be carbon neutral and hangs his wash on a clothesline to avoid powering a dryer. Elizabeth Kolbert, "Mr. Green: Environmentalism's Optimistic Guru Amory Lovins," *New Yorker*, January 22, 2007.
11. For this example, see Richard Hirsh, *Power Loss: The Origins of Deregulation and Restructuring in the American Utility System* (Cambridge, MA: MIT Press, 1999), 344 n. 2.
12. *World Energy Assessment* (New York: UNDP, 2000).
13. "Sustainable Development Knowledge Platform," at https://sdgs.un.org/#goal_section. The UN estimates that "in 2015 still about 2.8 billion people have no access to modern energy services and over 1.1 billion do not have electricity." In 2011, the Sustainable Energy for All initiative was created by the UN secretary-general with a primary objective of "providing universal energy access to modern energy services." https://sdgs.un.org/topics/energy#milestones.
14. See, for example, David McDonald, ed., *Electric Capitalism: Recolonizing Africa on the Power Grid* (Cape Town: HSRC Press, 2009); Ronen Shamir, *Current Flow: The Electrification of Palestine* (Stanford: Stanford University Press, 2010); Leo Coleman, *A Moral*

solutions for the developing world, such as microgrids.[15] But few challenge the central assumption of electrical modernity as a universal goal.[16] Modern capacities and freedoms have been so entirely conditioned by electrical infrastructure that the possibility of its failure is beyond imagination. The picture of a green grid allows one to imagine bringing the offgrid billion onto the grid, globalizing the electric modern without encountering a planetary limit.

For many who think of climate change as an issue of ethics and justice—whether religious, secular, or otherwise—technological solutions become suspect in just this way; they are seen as evading a fundamental ethical challenge. As Pope Francis put it in *Laudato Si*, "We have certain superficial mechanisms, but we cannot claim to have a sound ethics, a culture and spirituality genuinely capable of setting limits and teaching clear-minded self-restraint."[17] The pope, to be sure, also favored decarbonizing power grids, and indeed expanding them to the world's poor. "There is an urgent need to develop policies so that, in the next few years, the emission of carbon dioxide and other highly polluting gases can be drastically reduced, for example, substituting for fossil fuels and developing sources of renewable energy. Worldwide there is minimal access to clean and renewable energy." This call for expanded electrification and retooling would

Technology: Electrification as Political Ritual in New Delhi (Ithaca: Cornell University Press, 2017); Sunila Kale, *Electrifying India: Regional Political Economies of Development* (Stanford: Stanford University Press, 2014); Moses Chikowero, "Subalternating Currents: Electrification and Power Politics in Bulawayo, Colonial Zimbabwe, 1894–1939," *Journal of Southern African Studies* 33, no. 2 (June 2007): 287–306.

15. See, for example, the papers published by TERI (The Energy and Resources Institute, India), at https://www.teriin.org.

16. David McDonald, for example, begins his edited collection by noting that electrification furthers the Millenium Development Goals of fighting poverty and hunger, expanding education, promoting gender equality, reducing child mortality, improving maternal health, and combating HIV/AIDS as well as malaria and other diseases. *Electric Capitalism*, xvii.

17. https://www.vatican.va/content/francesco/en/encyclicals/documents/papa-francesco_20150524_enciclica-laudato-si.html.

seem to be one of the "superficial mechanisms" of which the pope spoke, insofar as it is a way of avoiding what he saw as the need for a more fundamental asceticism.

Laudato Si, rather unusually, calls our notice to this tension between the strategy and its ultimate goals, and it does so in language especially pertinent to the grid: "The technological paradigm has become so dominant that it would be difficult to do without its resources and even more difficult to utilize them without being dominated by their internal logic. It has become countercultural to choose a lifestyle whose goals are even partly independent of technology, of its costs and its power to globalize and make us all the same." With this remark, the pope retrieves a tradition of critical thought about infrastructure that motivated much of the environmental movement, at least from Thoreau to McKibben. Since then, though without much notice, the rise of grid-tied renewables in the age of climate change has made such thinking even more "countercultural" than it was in the 1960s. So it bears asking anew: In betting the future on a more distributed, more resilient, more networked system of renewable electric power, available to all, how might we also become dominated by its inner logic? Granting that *Electrify everything* seems both necessary and urgent, what exactly are we committing to, or being committed to?

"The most salient characteristic of infrastructure," Paul Edwards has written, "is the degree to which most technology is *not* salient, most of the time."[18] Indeed, this aperçu has been an almost ritual starting point for the recent wave of thought about infrastructure. Susan Leigh Star's influential essay "The Ethnography of Infrastructure" begins by saying, "This article is a call to study boring

18. Paul N. Edwards, "Infrastructure and Modernity: Force, Time, and Social Organization in the History of Sociotechnical Systems," in *Modernity and Technology* (Cambridge, MA: MIT Press, 2002), 185–225.

things. Many aspects of infrastructure are singularly unexciting."[19] And Lisa Parks begins her study of satellite dishes with the remark, "Most people know very little about the infrastructures that surround them in everyday life, whether electrical systems, sewer pipes, or broadcast networks."[20] Edwards, Star, and Parks all begin by noting something paradoxical about the project of knowing something that does not want to be known. John Durham Peters picks up this line of thought in *The Marvelous Cloud* and gives it the name "infrastructuralism." Peters suggests that twentieth-century thought was everywhere haunted by infrastructural transformation and its new forms of boredom and obviousness. "Freud's famous dictum, 'Wo es war, soll ich werden' (Where it was, I should be), might be understood as the imperative to make all infrastructures clear. There is a deep infrastructural ethic in modern thought."[21] And yet, as he notes, "Forgetting seems a key part of the way infrastructures work" (36).

It is not happenstance or laziness that keeps us from thinking about infrastructure. As Parks writes, "Not only are people socialized to be unaware of such systems; infrastructures are often designed purposefully to be invisible or transparent, integrated with the built environment, whether submerged underground, covered by ceilings and walls, or camouflaged as 'nature.'" She calls this "infrastructural concealment."[22] The electrical power system is probably the most pervasively concealed feature of our world. An architecture of invisibility shrouds it from view, reinforced by habits of disattention. Gretchen Bakke puts it well, writing that the grid

19. Susan Leigh Star, "The Ethnography of Infrastructure," *American Behavioral Scientist* 43, no. 3 (1999): 377–91.

20. Lisa Parks, "Technostruggles and the Satellite Dish: A Populist Approach to Infrastructure," in *Cultural Technologies: The Shaping of Culture in Media and Society*, ed. Göran Bolin (London: Routledge, 2012), 64–84.

21. John Durham Peters, *The Marvelous Cloud: Toward an Elemental Philosophy of Media* (Chicago: University of Chicago Press, 2015), 34–35.

22. Parks, "Technostruggles," 64, 66.

> is the world's largest machine and the twentieth century's greatest engineering achievement and we are remarkably oblivious to it. Experts at unseeing its wires and poles, we are equally unlikely to be able to differentiate a power station from an oil refinery. Substations, essential to our grid, flit perhaps at the edges of recognition, though most have been well hidden behind cement walls or stuck in out-of-the-way corners of our cities and towns. Transformers, once a world-altering technology, have also been rendered utterly unspectacular, secreted inside gray canisters that cluster like coconuts at the tops of urban utility poles.[23]

Wires run behind walls. Obviously. Could it be otherwise? Bakke's language here seems to invite the infrastructural ethic: To think about climate change requires us somehow to bring the grid and our dependence on it into consciousness, into shared concern, and into politics.

Susan Leigh Star goes farther by suggesting that "infrastructure is *by definition* invisible, part of the background of other kinds of work."[24] We never think of ourselves as using the grid; we just think we are using toasters and computers. For such reasons the infrastructure of the grid is environment as much as it is technology, "as ordinary and unremarkable to us," Edwards notes, "as trees, daylight, and dirt."[25] Even if we wanted to bring it into consciousness, where would we start? Bakke notes that the engineers and regulators responsible for it themselves have difficulty comprehending the whole, given its sheer scale and complexity. "As implausible as it must sound, the machine that holds the whole of our modern life in place 'works in practice, but not in theory.' No one can see, grasp, or plan for the whole of it. . . .

23. Bakke, *The Grid*, xii–xiii.
24. Star, "The Ethnography of Infrastructure," 380; my emphasis.
25. Edwards, "Infrastructure and Modernity," 185.

This is our grid in a nutshell: it is a complex just-in-time system for making, and almost instantaneously delivering, a standardized electrical current everywhere at once."[26]

The electric power grid is more than an object of forgetting; it is an engine of unconsciousness. The grid shares the invisibility of all infrastructure, but it also has features of its own that make it disappear. First among these is the organizing metaphor of "the grid" itself. Nicole Starosielski, in her excellent study of undersea fiberoptic cables, suggests that the habit of seeing infrastructure as a network or a grid is inherently dematerializing and deterritorializing, because of the primacy of abstract space in organizing it. "The environments that cables are laid through—the oceans, coastal landing points, and terrestrial routes—are seen as friction-free surfaces across which force is easily exerted, and where geographic barriers are leveled by telecommunications."[27] The idea of a network subtracts the need to be conscious of underlying geography. Power circulates, as far as most of us can tell, in abstract geometry. The grid has its own geography, but also its own anti-geography, so to speak; it is inherently deterritorializing. Now you see it, now you don't. One could imagine a map of all the oddly shaped parcels of land that have stanchions, transformers, or cables on them—the junked-up interstices between things we actually notice. But when one imagines the grid, it has less

26. Bakke, *The Grid*, xxvi, 7.

27. Nicole Starosielski, *The Undersea Network* (Durham: Duke University Press, 2015), 5. Elsewhere she writes: "Contemporary networks are often imagined as a distributed mesh, in which individual nodes are multiply linked in an amorphous and flexible topology. These distributed systems are not simply opposed to centralized structures, but, as Alexander Galloway has noted, the 'distributed network is the new citadel, the new army, the new power'" (11). "Analyses of twenty-first-century media culture have been characterized by a cultural imagination of dematerialization: immaterial information flows appear to make the environments they extend through fluid and matter less" (6). In this respect "media culture" may be shorthand for media that are in turn mediated by the grid.

to do with these actual spaces than with immaterial schematics and whole territories.

The metaphor of "the grid" is especially apt, from this point of view, because it is a two-dimensional schematic realized in three-dimensional material structures. It is an extreme case of abstract space superimposed on the built environment, a "concrete abstraction," as Henri Lefebvre put it.[28] For Starosielski, the result is a problem of knowledge; her own work is intended as its corrective, making "the physicality of the virtual" transparent. The cables, like other structures of the grid, are "invisible to the publics that use them," but her own study "reintroduces such a consciousness, one might even say an environmental consciousness, to the study of digital systems."[29]

A noble hope, certainly, and one that motivates this lecture as well. But the electric grid produces and mediates unconsciousness in ways that are not easily overcome by a lecture such as this. Of course it is possible to remind ourselves that both the grid and its electrical current are material phenomena in real space. But they are designed to be experienced otherwise. The perspective of the user—more accurately, the nonperspective of the "user"—is that power is just there, always. We are oblivious not because we haven't happened to notice the materiality of the grid, and not just because of the scale of infrastructure; the grid produces an unconsciousness that one experiences, paradoxically, as freedom from care. From the earliest emergence of electric power systems, even before they were articulated as "the grid," what has been supplied *as power* has been designed so that its subjective content would be the assumption of constant

28. Henri Lefebvre, *The Production of Space* (1974), trans. Donald Nicholson Smith (Oxford: Blackwell, 1991), 86.

29. Starosielski, *The Undersea Network*, 15, 3. See also the similar project in Andrew Needham's study of the power system in regional Phoenix (including the Navajo Nation), *Power Lines: Phoenix and the Making of the Modern Southwest* (Princeton: Princeton University Press, 2014).

and unlimited capacity of electricity as a generalized medium. The constancy of the grid is, for users, the constancy of their own ability to do things that involve electric power, at will and whenever they like. We actively dislike being called out of this empowering sleep—for example, by having to see a substation or a wind farm. We feel entitled to our obliviousness. To an environmental tradition that has always emphasized awareness and care, this is the scandal: The electrical power grid, like the other utility infrastructures that preceded it, is designed to mediate *an entitled obliviousness to where our resources come from and where our waste goes*—a commitment no one is aware of having.

WHAT IS THE GRID, AND HOW DID IT GET THAT WAY?

People who engineer, operate, or regulate the grid know that it is not, so to speak, a thing: There are transmission grids and distribution grids; systems can be radial, networked, looped, or tied ring grids; they can operate on synchronous AC or variable frequency; increasingly they incorporate microgrids and other modifications; they frequently cross borders even when they are called "national grids." The metaphor of the grid allows this vast field of connection to be perceived as a unity. It is a radical simplification. One of the reasons for the continued use of the singular, "the grid," is that all existing versions of it are designed so that you, as *user*, do not have to know any of this.

Many of the features that make it natural to speak of electrical infrastructure as "the grid" were in place well before Edison built the first central electric station in 1882; others did not develop until long after. Electrical delivery systems did not come to be known as *the grid* or *the gridiron* until the 1920s. The UK, for example, passed the

Electricity Act in 1926, which authorized the creation of a "national gridiron." The metaphor came into play in part because of a change in electrical infrastructure design and its governmental frame after World War I. Earlier electric plants worked on a hub-and-spoke model with a central plant radiating power outward to nearby consumers. At first these were small and local; Edison's famous Pearl Street Station supplied DC current and could only reach a few city blocks in any direction. Most generating systems were self-contained; in 1884 there were eighteen power stations on Edison's model, but 378 electrical generating stations that belonged to factories, hotels, and rich proprietors like J. P. Morgan. The largest of these early electric systems were tiny by our standards; they would now be thought of as microgrids. They catered to the wealthy; there was little thought as yet of a grid for the masses. Moreover, hub-and-spoke power systems were distinctly urban infrastructure. They were initially designed for one purpose, which they took over from gas: light. They distinguished the now brilliantly lit cityscape from its darker surrounds. This began to change in the interwar period as governments began to push for rural electricity. Over the succeeding century the deindustrialization of the city and the industrialization of rural sectors (principally agriculture) collapsed this electric polarity of country and city that was so pronounced in the industrial era. The shift to AC expanded these circuits rapidly because it allowed central plants to send electricity for greater distances (stepping up and stepping down at substation transformers), as when the hydro plant at Niagara Falls was built in 1895 to power Buffalo, twenty-six miles away. At the same time, the spread of AC motors eventually allowed electricity to become a generalized medium of power—domestic and industrial alike—rather than just a lighting technique.

In the United States in the 1920s the hub-and-spoke systems began to be connected to one another, partly because the federal government wanted to reduce the power outages that had become a problem during World War I. Louis Hunter and Lynwood Bryant note that "the term *interconnection* for this phenomenon was first brought to public attention during the power emergency in the last year of the war, when it was applied to the linking together by means of high-tension lines of two or more independently operated central-station systems."[30] Interconnected power systems provided a protective layer of redundancy, created economies of scale, pooled reserve capacity, and allowed more flexible demand management. "As early as 1922 the power systems of Massachusetts, New Hampshire, and Rhode Island were said to be so thoroughly interconnected that they could be regarded as a unit for purposes of power supply."[31] Electric power, in short, had become a network. In the process, the transmission grid came to be distinct from the production of power. The Rural Electrification Administration, for example, "helped to establish distribution networks and left most production in private hands."[32] Power from coal, hydro, oil, and other sources mingled on the same transmission and distribution grids, which covered a hodgepodge of territories in ways that remained invisible to end users (then as now). The period between the wars, then, was the critical period in which electric power systems became grids, calling into being new kinds of state regulation while also coming to represent social modernization and freedom.

Even before the language of the power grid acquired salience from the interconnections of the "gridiron," however, electric power transmission had already been developed by means of conceptual frames that drew on earlier grids. So it is worth a brief detour through

30. Louis C. Hunter and Lynwood Bryant, *A History of Industrial Power in the United States, 1780–1930*, vol. 3: *The Transmission of Power* (Cambridge, MA: MIT Press, 1991), 365.

31. Hunter and Bryant, *History of Industrial Power*, 366.

32. David E. Nye, *Electrifying America: Social Meanings of a New Technology* (Cambridge, MA: MIT Press, 1990), 314.

the nineteenth-century history that made the grid, when it came, seem natural.

Edison often said that his goal, even before deciding that incandescent bulbs would be part of the system, was "the indefinite subdivision of light." He told the *New York Sun* in October 1878 that he had seen how new electric generators had enabled transmission at a distance, but he "saw that what had been done had never been made practically useful. The intense light had not been subdivided so that it could be brought into private houses."[33] The incandescent bulb was only one detail in this larger process, which involved grid logic at several stages: first, the arrangement of the wires into parallel circuits, so that when one bulb went out it would not stop flow to other bulbs; later, by the same principle, a feeder-and-main distribution network, nesting circuits within circuits so that individual houses would be connected only to larger circuits, not each other.[34] Bakke notes that "the very existence of a relatively dim bulb, which we take for granted today, was made possible only by the prior invention of the parallel circuit."[35]

It hardly seems accidental that Edison developed the basic schema of an electric grid in the city that had already become notorious—for better and worse—for the radical application of a gridiron design to its landscape. One of his patent applications, submitted in 1887 for a "System of Electrical Distribution," shows in each of its four illustrations the variations Edison was working on grid concepts. One

33. Quoted in Paul Israel, *Edison: A Life of Invention* (New York: John Wiley and Sons, 1998), 166.

34. Jill Jonnes, *Empires of Light: Edison, Tesla, Westinghouse, and the Race to Electrify the World* (New York: Random House, 2004), 67–68. The incandescent bulb, as many historians note, was not even a new idea. Some give priority to the British inventor Joseph Swan, and experiments in the field had been made by others for decades. See Thomas Hughes, *Networks of Power: Electrification in Western Society, 1880–1930* (Baltimore: Johns Hopkins University Press, 1983), 21–32.

35. Bakke, *The Grid*, 31.

of them also shows how the grid of wires would be superimposed on the Manhattan street plan. The area covered by the network of the Pearl Street Station had been laid out before the famous 1811 plan for Manhattan's streets, but it was already a grid: fifty-one blocks bounded by the East River, Spruce, Wall, and Nassau Streets.

The abstractive icon of the urban grid has a long history of association with state power, stretching from Babylon to the Roman, Chinese, and early modern Spanish empires. New Haven was one of the first examples in English North America (1638); others include Philadelphia (1682) and Savannah (1733). Grid planning became national after the Revolution, in the Public Land Survey System (1785), one of the few major actions under the Articles of Confederation (updated in the Northwest Ordinance two years later); it laid out territories such as Wisconsin on a grid in order to propertize them for white settlement. The Manhattan grid of 1811, too, was designed to facilitate the sale of common land holdings on the real estate market; it was speculatively open-ended, commensurating space into property.[36] The indefinitely extensible grid invited people to imagine that all places could be organized by abstract space—provided one could look past the practical cost of superimposing grid reality on landscapes and infrastructures. Manhattan's grid plan looks

36. Ted Steinberg, *Gotham Unbound: The Ecological History of Greater New York* (New York: Simon and Schuster, 2014), 41–65. Parts of Manhattan had been turned into grids well before northward development required the 1811 plan. When Nicholas Bayard sold his hundred-acre farm in 1788, for example, he divided it into a grid; present-day Houston Street descends from this moment. See Gerard Koeppel, *City on a Grid: How New York Became New York* (New York: Da Capo Press, 2015). Koeppel also records a long history of critiques of the grid, beginning at least with Clement Clarke Moore in 1818. "Our public authorities seem unwilling to depart from their leveling propensities, but proceed to cut up and tear down the face of the earth without the least remorse, and, apparently, with no higher notions of beauty and elegance than straight lines and flat surfaces placed at angles with the horizon, just sufficient to suffer the mud and water to creep quietly down their declivities." Moore himself later cashed out on the grid by converting his estate into the gridded development now known as Chelsea (Koeppel, 136–38).

intuitive to millions of strangers every year partly because the landscape has been blasted, filled, and leveled for centuries, in a process that began in the seventeenth century but achieved peak violence during the ascendency of the grid and the first reactive wave against it: the years from Casimir Goerck and James Randel, who surveyed the grid, to Frederick Law Olmsted, who designed Central Park as an escape from it.[37] These material dimensions of grid architectures, like the wires and pipes and transformers of the power grid, are designed to disappear by conforming to abstract space. Edison, who buried his wires beneath the streets of the urban grid, counted on their invisibility to the end user. To be sure, he also counted on a constant stream of horse-carts hauling coal to the smoky furnaces of the central stations.[38] The grid, that is to say, acquired its immateriality over time, but the utopia of its spaceless pervasiveness was already prepared by the streets.

One enabling condition of this illusion is that all grid media are insulated. We are careful not to touch them, for good reason if we are on an AC grid. Before Edison, electricity was powerfully tied to the body's own conductivity, and in the eighteenth century and early nineteenth century the demonstration of electricity frequently involved feeling it; there was a strong connection between electricity

37. Hilary Ballon, ed., *The Greatest Grid: The Master Plan of Manhattan, 1811–2011* (New York: Museum of the City of New York and Columbia University Press, 2012); Richard Sennett, "American Cities: The Grid Plan and the Protestant Ethic," *International Social Science Journal* 42, no. 3 (1990): 269–85; Alan Trachtenberg, "The Rainbow and the Grid," *American Quarterly* 16, no. 1 (1964): 3–19; Peter Marcuse, "The Grid as City Plan: New York City and Laissez Faire Planning in the Nineteenth Century," *Planning Perspectives* 2 (1987): 287–310. For a contrasting view of the grid, see Rem Koolhaas, *Delirious New York: A Retroactive Manifesto for Manhattan* (New York: Oxford University Press, 1978).

38. When electric power was run to J. P. Morgan's house, in one of Edison's first high-profile demonstrations, its steam-powered generator sat next to the house, belching smoke from its coal furnace in the hours when the engineer was on site to run it. It could hardly have been absent from anyone's mind inside the house, even after it was bricked up to subdue the noise.

and various projects of mesmerism, medicine, and spiritualism. By contrast, the constant hazard of electrocution in grid media means that we only ever touch insulated surfaces. The counterimages of the electric chair and shock therapy make us dimly but mortally mindful that the grid lies beyond our contact. Thus we are constantly encouraged to think of grid media as immaterial, and perhaps this insulation of the haptic helps explain why grid media occupy only abstract space. It is not that they are intangible; we touch their insulated surfaces all the time. But this touch is oriented to what it cannot feel, making grid media appear to be less tactile than other media of inscription: stone, clay, animal skin, paper, and so on.[39]

The apparent intangibility of the grid was driven home in the grisliest episode of Edison's early years. Hoping to discredit the AC systems of his rival George Westinghouse, Edison contrived to have Auburn prison employ AC equipment made by Westinghouse for the first penal electrocution, in 1890—a repellent and botched business that drove participants out of the room vomiting. He even tried to put "Westinghouse" into circulation as a verb for "electrocute." Dogs, calves, horses, and a Coney Island elephant named Topsy were all fried to death as part of Edison's public campaign to stop the shift to AC power.[40] He failed in that aim, but the conductivity of the body (Whitman's "body electric") would never be experienced the same way again.

The neutralization of space by power transmission had another clear analogue in the early period of electrification: the telegraph. At the same time that he was thinking about illuminated space as an urban grid, Edison was drawing on his familiarity with the telegraph

39. Compare McLuhan's certainty that "all media are extensions of our own bodies and senses." Marshall McLuhan, *Understanding Media: The Extensions of Man* (1964) (Cambridge, MA: MIT Press, 1994), 116.

40. Richard Moran, *Executioner's Current: Thomas Edison, George Westinghouse, and the Invention of the Electric Chair* (New York: Vintage, 2003).

as a means of transmission at a distance. He had begun as a telegraph operator, and his earliest inventions were improvements to the telegraph. Many of his designs for the electric grid imported techniques with which he was already familiar, such as parallel circuits and electromagnetic relays. Naturally, when it came time to lay the wiring for his first trial power station, he brought in workers from Western Union.[41]

The telegraph, as James Carey showed in a classic essay, was "the first electrical engineering technology and therefore the first to focus on the central problem in modern engineering: the economy of a signal."[42] It was the first infrastructure to separate communication from transportation, thus transforming the media concept from a generalized conception of means to the specifically modern idea of technically enabled communication at a distance.[43] Edison's power station was a medium of electricity in this sense; and the grid is the medium of all media that have since been devised. With only a little forcing we could call it the common premise of media as such—print and manuscript having been retroimagined, after McLuhan, in light of electronic media.[44]

41. Jonnes, *Empires of Light*, 64.

42. James W. Carey, "Technology and Ideology: The Case of the Telegraph," in his *Communication as Culture: Essays on Media and Society* (New York: Routledge, 1989). My thanks to John Durham Peters for this reference and for sharing his own essay about it, "Technology and Ideology Revisited" (ms.). "Carey's telegraph," Peters notes, "though shrouded by flowery religious rhetoric at first, is an agent of the modern secular order, whose logic of space and time is homogeneous and whose empire is the grid."

43. On the deeper history here, see John Guillory, "Genesis of the Media Concept," *Critical Inquiry* 36, no. 2 (2010): 321–62.

44. McLuhan only barely noticed the electric grid, and when he did he mistook it for electricity itself. "Obsession with the older patterns of mechanical, one-way expansion from centers to margins is no longer relevant to our electric world. Electricity does not centralize, but decentralizes. It is like the difference between a railway system and an electric grid system: the one requires railheads and big urban centers. Electric power, equally available in the farmhouse and the Executive suite, permits any place to be a center, and does not require large aggregations." *Understanding Media*, 36. Elsewhere he writes, inaccurately, "It

Telegraphy had another bequest to the power grid: the social form of monopoly capitalism, which laid the pattern for the grid's relation to users. "The speed and volume of transactions demanded a new form of organization of essentially impersonal relations—that is, relations not among known persons but among buyers and sellers whose only relation was mediated through an organization and a structure of management."[45] Users are mediated not just by wires, in other words, but by utilities.

Utilities have their own arts of governance with their own techniques. Many of these, like demand management and the load curve, are normalizing in the strict sense of the term. Load management is an essential function of any power grid; demand comes in waves, so either more power has to be readied to meet it or excess power needs to be brought in from elsewhere on the grid. The premise of load management is too obvious to need stating: constant availability. The grid is managed so that users at peak demand will have all the power they want, just like any other time. This is not an attitude of consumerism on the part of the users—though it may be that too; it is the regulatory activity for which grids are designed and which is necessary to their operation. Colossal waste is necessary to this guarantee of constancy; the amount of power that has to be generated is many times greater than the amount actually used.[46] All the ways people are using power in the territory of the grid must be normalized around an imagined user of maximal demand; power is delivered not for a need but as a constancy, and thus people do not even need to think

is to the railroad that the American city owes its abstract grid layout, and the nonorganic separation of production, consumption, and residence" (104).

45. Carey, "Technology and Ideology," 205.

46. Amory Lovins calculates that in 2010, electricity generation accounted for 40 percent of that year's carbon emissions. "Two-thirds of this primary fuel on average—nearly half even in the most efficient power stations—is discarded as waste heat or used internally before the electricity leaves the power plant" (*Reinventing Fire*, 172). That doesn't count grid loss or the overproduction for voltage maintenance.

of themselves as having a demand for power. They just flip switches. Normal unconsciousness is the end product of the grid, and the vast majority of generated power is spent keeping it that way.

Power has come to be imagined as constantly on demand in abstract space, addressing an indefinite number of people as sometimes actual and always potential users. We might say that an indispensable component of any grid is this abstraction: *the user*, a private subject of maximal demand and minimal awareness. For as long as utility grids have been around this structure has been integral to their design. The metered switch between you and the grid converts what you do, no matter how your actions are motivated, into quantified demand or "load." Economic incentives of demand management, like pricing, have from this point of view the same output as any normatively charged awareness of the planet; insulation in your attic might make more difference than all of your good (or ill) will toward the earth or its people. On the grid as it is currently designed, the meter is the only means by which you are made aware of your power. If you look at it at all, your actions are translated back to you as usage alone. (This is true even of what are called "smart meters.")

Even more than telegraphy, the model utility for the beginning of electrification was gas. Competing with domestic gas lighting was Edison's primary motive, and both the distribution and the pricing of incandescent light were designed accordingly, often dictating choices in the engineering.[47] Edison adopted the usage meter from gas infrastructure and ran wires through pipes that had been laid for gas. George Westinghouse—who took Edison's scheme for a DC station and built the first AC system, using high voltage by means of transformers and substations to send power by much greater distances—turned to electricity only after venturing a similar scheme for the delivery of natural gas: it stepped up pressure through narrow

47. Nye, *Electrifying America*, 165; see also Hughes, *Networks of Power*.

pipes so as to reach greater distance before being lowered to a level for home use. (One historian notes that all of Westinghouse's early ventures—air brakes, railroad signals, gas pipes, and the AC grid—involved transmission over distance. "Many of the technologies that George Westinghouse pursued involved a crucial linking mechanism that served to connect the long-distance transmission lines with the rest of the system. Often these devices incorporated feedback mechanisms that regulated the system.")[48]

The gas grid and electric power have conjoined histories, and in most parts of the United States they continue to be administered by the same utility companies. The earliest of these, in Baltimore, began supplying artificial gas from coal in 1816. By 1828 Broadway was brilliantly illuminated with gas flares, and by the time of the Civil War 183 urban gas lighting companies were operating in the United States.[49] Gas lighting enabled the nocturnal city even before its electrification, and thus gave Edison a preexisting grid to electrify, already populated with users. Gas utilities created the lived context for an expectation of constant supply on demand from a central, invisible, corporate source.

Even gas is not the beginning of this story. When the first gas company began operating in England in 1810, it was modeled in turn on the recent rise of water companies.[50] In Manhattan a water company that started running wooden pipes under the streets was directly involved with grid planning before the 1811 plan.[51] Before

48. Steven W. Usselman, "From Novelty to Utility: George Westinghouse and the Business of Innovation During the Age of Edison," *Business History Review* 66, no. 2 (1992): 267. See also Jonnes, *Empires of Light*, 129–30.

49. Nye, *Electrifying America*, 95. "Natural gas, first used to light several buildings in Fredonia, New York, in 1821, was long ignored as an energy source because of the difficulty of moving it to the consumer" (Nye, 121).

50. Wolfgang Schivelbusch, *Disenchanted Night: The Industrialization of Light in the Nineteenth Century*, trans. Angela Davies (Berkeley: University of California Press, 1995).

51. Koeppel, *City on a Grid.*

electric wires and gas jets came inside the home, water and sewer engineers and corporations had begun carrying resources into and waste out of cities by means of pipes, pumping stations, aqueducts, faucets, and drains.[52] The unconscious end user, defined by continuous potential of demand and distance from resources, was thus built by successive layers into the techno-social environment.[53]

Today of course it is the internet that most thoroughly organizes life by means of grid infrastructure. An offshoot of the grid that now helps to regulate the grid on which it depends, the internet has absorbed practically all media into grid mediation.[54] As with other grid-mediated media, the internet's grid stands between users and fuel sources. (How were you to know that Facebook has committed to renewables for its servers, but Google has not?)[55] The internet's relation to the grid that is its ground has become complex, especially given the rising language of securitization. For my purposes a key point is that the internet evolved in large part thanks to its own construction of the user.[56] As Tung-Hui Hu argues, "the time-shared user as an economic subject" has been naturalized into the experience

52. Joanne Abel Goldman, *Building New York's Sewers: Developing Mechanisms of Urban Management* (West Lafayette: Purdue University Press, 1997).

53. David Nye, *Consuming Power: A Social History of American Energies* (Cambridge, MA: MIT Press, 1998). "Overall, the adoption of networked technologies had an unintended collective effect: enmeshed in systems, families consumed energy—often without thinking about it—whenever they turned a tap, threw a switch, twisted a valve, or lifted a telephone receiver. The purchasing of goods and services was built into the very architecture" (Nye, 96).

54. "As recently as the year 2000, 75% of all the information stored by human societies worldwide was in analog formats such as paper documents, pictures, books, tapes, and X-ray films. By 2007, 94% of all stored information was in digital electronic form." Lovins, *Reinventing Fire*, 166.

55. Unless you were to read the Greenpeace report titled *How Green Is Your Cloud?*

56. Paul E. Ceruzzi, *A History of Modern Computing*, 2nd ed. (Cambridge, MA: MIT Press, 2003); Wendy Hui Kyong Chun, *Programmed Visions: Software and Memory* (Cambridge, MA: MIT Press, 2011); David Golumbia, *The Cultural Logic of Computation* (Cambridge, MA: Harvard University Press, 2009), 184, 188; Tarleton Gillespie, "The Stories Digital Tools Tell," in *New Media: Theories and Practices of Digitextuality*, ed. John Caldwell and Anna Everett (New York: Routledge, 2003), 107–26.

of available computing as a mode of private freedom.[57] Computing power in the age of the cloud increasingly presents itself as a utility, like the electric power it runs on; it makes possible a private user "with no concern for how many ticks of the processor one [is] using."[58] The cloud, argues Hu, "has replaced television as the premier mechanism for sorting the public into private users" (41). In other words, the internet borrows and extends the infrastructure of power it is built on, and many of the features people associate with the internet—its realization of abstract space, decentralized production, long-distance transmission by networked switching, a complex relation to national infrastructural politics, an aggressive colonization of other forms of life, modes of privatization, an assumed narrative of technological momentum, the creation of new elites of engineers, regulators, and capital—are features of the power grid, newly magnified.

Infrastructure, as Edwards puts it, is techno-social. In the case of the grid, the interarticulated technologies and their social dimensions have evolved into a bewildering complexity. As historians such as Thomas Hughes and David Nye have shown, the rapid expansion of electric power transformed almost every dimension of social life, from industry to domestic space, through an immense succession of electric inventions and the new patterns that arose around their adoption. Each of these innovations in electrical modernity, from cinema to the internet, has been mediated by the grid. Nightlife, air conditioning, and digital humanities are all new reasons to need electric power. For a century now, the grid has been the modern; and for many of those who don't have it, connection to the grid still stands for belonging in the modern world.[59] The historical process

57. Tung-Hui Hu, *A Prehistory of the Cloud* (Cambridge, MA: MIT Press, 2015), 39.
58. A phrase of Ceruzzi's (*History of Modern Computing*, 208) that Hu cites (*Prehistory of the Cloud*, 39).
59. Nye, *Electrifying America*, especially chapter 8, "The Electric Future."

behind this perception, however, was neither automatic nor simple. The historian Ronald Tobey shows that electrical modernization was rather limited before the New Deal: Building contractors before the 1930s, for example, routinely wired houses for illumination but seldom provided outlets to take a full range of electrical appliances. Americans did not rush to embrace electrical appliances the way they did for automobiles. The first major exception was radio; radio consoles built to plug directly into house current were marketed in 1925, and in short order the radio "broke the long-standing resistance of domestic consumers to high consumption of electricity."[60] It was New Deal progressives, including Roosevelt himself, who envisioned electrification as a democratic social modernization.

This phase of power history is notable for the way the coming of the grid transformed governmentality. Nations have taken different paths in building power systems, but all have claimed ultimate say and responsibility for a power grid. Variations in administrative and economic form notwithstanding, what is remarkable here is that within a relatively short period grid provision and regulation became a globally distributed expectation for government. When the grid fails, as it frequently does in developing countries, the state itself is felt to fail.[61]

60. Ronald Tobey, *Technology as Freedom: The New Deal and the Electrical Modernization of the American Home* (Berkeley: University of California Press, 1996), 22–23. For a vivid example of electrical progressivism, see David Lilienthal, *TVA: Democracy on the March* (New York: Harper, 1953).

61. "The power shortages and blackouts have cast a harsh light on elected officials, causing rising anger among voters for whom reliable electricity was supposed to be a dividend of democracy and economic growth. . . . 'It's not only a symbol of failure when the lights go off,' said Anton Eberhard, an energy expert and a professor of management at the University of Cape Town. 'It's experienced directly by people. If you're about to cook or if your child is studying for an exam the next day and your lights go off, people feel this very directly. There is a very concrete and dramatic expression of failure.'" Norimitsu Onishi, "Weak Power Grids in Africa Stunt Economies and Fire Up Tempers," *New York Times*, July 2, 2015. Jacques Ellul noted the reasons in *The Technological Society*, in the only paragraph he devoted to electrical power: "Electrical networks may remain for some time independent

Grid governmentality is to a significant degree independent from other forms of politics and ideology. Consider the vast literature on the differences between capitalism and socialism; then look around for a comparable literature on grid construction and maintenance as a *dispositif* of government. Cold War narratives disguise from our view the remarkable convergence of infrastructural development. All modern states—democratic or authoritarian, capitalist or socialist, former empires or former colonies—are committed to it. Lenin launched a national grid in Russia shortly after the Revolution, declaring that full socialism required the electrification of the Soviet Union.[62] Indeed, he went further: "Communism," he declared, "is Soviet power plus the electrification of the whole country."[63] Some European nations, like Germany, also produced power directly through state agency. The European powers began to set up grids in their colonies, leaving complicated legacies of "modernization" to the postcolonial nations.[64]

of one another. But this situation cannot last when it is found that independence gives rise to general costs of no inconsiderable magnitude, difficulties in arranging the courses of the lines, and even practical difficulties in electrical technique. The interconnection of electrical networks is demanded by all technical men [*sic*]. Again, the only question is: who will execute it? And it is immediately clear that only the state is in a position to do so." Jacques Ellul, *The Technological Society* (1954) trans. John Wilkinson (New York: Vintage, 1964), 237.

62. Nye, *Electrifying America*, 138.
63. Lewis Mumford gives this a milder translation: "Electrification plus socialism equals communism" (*Technics and Civilization* [New York: Harcourt Brace Jovanovich, 1934], 264). V. I. Lenin, "Our Foreign and Domestic Position and Party Tasks: Speech Delivered to the Moscow Gubernia Conference of the R.C.P.(B.), November 21, 1920; published in 1920 in the pamphlet *Current Questions of the Party's Present Work*, trans. Julius Katzer, in *Collected Works*, 4th ed, vol. 31 (Moscow: Progress Publishers, 1965), 408–26. Online at https://www.marxists.org/archive/lenin/works/1920/nov/21.htm.
64. See, for example, Y. Srinivasa Rao, "Electricity, Politics and Regional Economic Imbalance in Madras Presidency, 1900–1947," *Economic and Political Weekly* 45, no. 23 (2010): 59–66; and Moses Chikowero, "Subalternating Currents: Electrification and Power Politics in Bulawayo, Colonial Zimbabwe, 1894–1939," *Journal of Southern African Studies* 33, no. 2 (2007): 287–306.

In countries like the United States and most of Europe, electrical power first developed by means of competitive private capital and a rapid proliferation of power systems. As a check to this somewhat chaotic development, local governments often got directly involved. In 1923, there were 3,083 municipalities in the United States with their own electrical systems.[65] Public, private, and cooperative utilities eventually consolidated under the legal doctrine of the "natural monopoly." In 1934, partly because of the financial collapse of Samuel Insull's web of electric companies, and partly to foster interconnection, the Public Utility Holding Company Act recognized electric companies as "public goods," bringing about a new regime of utility regulation. Utilities had argued that it was both inefficient and disruptive to tear up the streets for two or more parallel sets of transmission lines or pipes. "According to this line of thinking," notes David Nye, "the networked city by its very nature required but one water system, one telephone company, one electrical utility, one gas company, and so forth. Energy monopoly was 'naturalized' as a condition for doing business."[66] Since 1978, the corresponding monopsony power of utilities as energy buyers has been broken; power can be produced and distributed by multiple firms, or even private net-metered homes.[67] But all still use a common transmission grid.[68]

As technical developments turned electricity into larger and larger systems, economies of scale gradually shifted the focus of this regulation from local to regional jurisdictions. Early in the process, Samuel Insull realized that power station design, because it was oriented to transmission at a distance, would have greater scope in the

65. Tobey, *Technology as Freedom*, 45–48.
66. Nye, *Electrifying America*, 124.
67. Hirsh, *Power Loss*.
68. On the various regulatory environments for the early development of power grids, see Hughes, *Networks of Power*; and Hirsh, *Power Loss*.

jurisdiction of the state of Illinois than in the city of Chicago.[69] He lobbied for the creation of state public utility commissions, which spread rapidly after 1907.[70] The grid, however, soon moved beyond state lines; new knowledges and jurisdictions were necessary to the era of "Giant Power," as Pennsylvania governor Gifford Pinchot called it in 1924. Grid regulation gave rise to "regional planning," carried out by new cadres of engineering and policy experts.[71] In this way the grid seemed to demand the reconfiguration of the state. Access to electric power came in the same years to define the modernity of life, and no government has been able to avoid it for long.

THE ETHICS OF THE GRID?

The infrastructure of power in our society creates conditions in which power use is intractable to ethical life as it is usually understood. As Bernard Williams observed, most modern conceptions of ethics imagine a realm of individual choices, obligations, or dispositions that become salient in opposition to desire. Ethical being of this kind would gain traction on climate change if we imagined reducing our carbon footprint as a matter of using less power, of reining in our demand through greater humility, or thrift, or stewardship of the earth, or fairness to the world's poor, or care for the people of the future.

Decarbonization is difficult to imagine in this register for a variety of reasons. The greening of the grid has become an engineering task and a regulatory project precisely so as not to require anyone to curb

69. Thomas Parke Hughes, *American Genesis: A Century of Invention and Technological Enthusiasm, 1870–1970* (New York: Viking, 1989), 230.
70. Lovins, *Reinventing Fire*, 173.
71. Hughes, *American Genesis*, 356–57; "Giant Power," *The Survey: Graphic Number* 51, no. 11 (March 1, 1924); and "The Regional Community," *The Survey* 54, no. 3 (May 1, 1925).

a desire for energy. Energy use, meanwhile, has become a condition of belonging to the modern. As utility users, we find that using power entails no consciousness of resources. What it does entail is a special sense of freedom, of capacities undefined by particular ends. This is the ethical subjectivity (in Foucault's sense) of the grid. The standing reserve of power that surrounds us, our capacity to do we know not what, structures our subjectivity and self-relation in very particular although mostly unrecognized ways. Obliviousness to these conditions is normal life, to the point that wresting ourselves outside of it would require a countercultural mode of life burdened by efforts that would seem, to most, irrational. For these reasons, the obliviousness that is our normal relation to the grid, when forced into consciousness, translates to flat affects of indifference, or powerlessness, or worse.

As others have shown, ethics in the context of climate change can mean a tremendous range of things, mostly beyond my scope here: not just what Bernard Williams called "the morality system," or even the virtues and dispositions of concern for the planet that Dale Jamieson so eloquently describes, but complex issues of responsibility for risk, what it means to care for future generations or for nonhuman life, the ethics of design and invention in the technosphere, fairness and carbon pricing, and so on.

Many of the ethicalizing programs I have just named have problems of their own. Stewardship, for example, is easiest to imagine as an ethic of property owners; Aldo Leopold was very candid about the way his "land ethic" was rooted in his landowner's sense of responsibility to his farm in Wisconsin. This is not nothing, and in fact I share it; but private property is the very opposite of a commons, and in its own way is like infrastructure in that it disposes obligations and freedoms within a very limited sphere demarcated by a system of enclosure that no property owner can materially alter. Virtues imagined as restraints on desire are open to the suspicion of being crypto-Protestant (perhaps why they work best in Nordic countries like Lutheran Denmark) or the distinctive sensibilities of educated elites. Moreover, although most ethical philosophy seeks to clarify norms that would hold universally—if only we could describe the reasons well enough—any such sensibility that one might articulate enters immediately in the realm of political differentiation. Think of the way the image of the Prius-driving liberal so quickly became an article of populist resentment, and thus fueled proud and defiant gas-guzzling in other quarters. Ethicalizing programs tend to ignore such dynamics because they project an efficacy to individual obligation by imagining that remoralized dispositions will aggregate, in the mode of a social movement. (A similar imagination seems to lie behind academic enthusiasm for metaphysical redescription of the posthumanist kind; the utopian hope depends on the idea that the academic project is a pioneering instance of what we expect to become, by some process or another, a universal sensibility.)

Infrastructural solutions to greenhouse gas emissions cut past these limitations because they affect everyone, independently of their conscious commitments and differences. They can be applied in authoritarian cultures as easily as in democratic ones, if not more so. And to the extent that they can be harnessed to the performative power of narratives of inevitability, they are freed from much of the burden of recruiting political will.

The grid, in short, is a problem for those traditions of environmental ethics, inherited from the nineteenth and twentieth centuries—that is to say, before climate change became the central issue—that construe ethics as awareness and agency in the face of obligations that are universally salient. The grid so buffers us from fuel sources that changing our consumption of power has less traction on carbon emissions the greener the grid becomes; the long-standing suspicion of "consumerism" no longer gets at the issue of carbon emissions. The grid, moreover, produces an unconsciousness that we experience only indirectly, as freedom. It involves us in distributed and normalized agency that scarcely reflects whatever might be our own motives. It construes us as users related to one another through utilities; as bearers of constant potential demand for unspecified ends; as subjects of comfort and convenience; as private consumers located in history by narratives of apparently inevitable technological shift; and as citizens whose only connection to the governance structures of grid modernity is that of rate payers. With the growing near-consensus of a green grid, then, the long history of attempts to call us back to awareness and responsible action finds itself at a limit.

And yet the rise of environmental awareness in modern societies has been linked from the start to a sense of ethical challenge. The more holistic the perception, the more countercultural it is understood to be. In what follows, I review some key moments in the dialectical formation of ethics against the grid, in order to sharpen a sense of what has changed.

Lecture II

Off the Grid

MICHAEL WARNER

The utopia of limitless green energy is older than the concept of energy itself, and older than the use of petroleum. "Energy" as an abstraction, and with it the science of energy, only arose once the conversion of energy sources had become a practical issue. As British capitalists increasingly turned from water power to coal-fired steam, they motivated a new set of reflections on energy equivalents, leading to the formulation of the laws of thermodynamics that are now the foundations of energy science. But this happened quite late in the process: William Rankine's formulation of the first two laws of thermodynamics, for example, dates from the early 1850s, and his speculative paper "Outlines of a Science of Energetics" was first delivered in 1855.[1]

1. Crosbie Smith, *The Science of Energy: A Cultural History of Energy Physics in Victorian Britain* (Chicago: University of Chicago Press, 1998). See also the important early paper by Thomas Kuhn, "Energy Conservation as an Example of Simultaneous Discovery," which makes the connection between the emergence of energy science and the prior context of conversion processes and practical machinery (1959; repr. in Kuhn, *The Essential Tension* [Chicago: University of Chicago Press, 1979]). See as well Cara Daggett, *The Birth of Energy: Fossil Fuels, Thermodynamics, and the Politics of Work* (Durham: Duke University Press, 2019), and Anson Rabinbach, *The Human Motor: Energy, Fatigue, and the Origins of Modernity* (Berkeley: University of California Press, 1992).

On the Grid. Michael Warner, Oxford University Press. © Regents of the University of California 2025.
DOI: 10.1093/oso/9780197696248.003.0004

Two decades earlier, in 1833, a German immigrant named John Adolphus Etzler published a heady survey of power sources under the title *The Paradise Within Reach of All Men, Without Labor, by Powers of Nature and Machinery*. Etzler, who came to America in pursuit of development schemes with his partner Washington Roebling—later the designer of the Brooklyn Bridge—was a utopian of the baldest variety, as his title suggests. Unlike Roebling, he was also a terrible engineer, and the machines he imagined turned out to be hopelessly impractical. But his basic vision remains of interest because he was principally excited by the prospects of wind, tidal, and solar power, all of which he treats as variants of the single medium of power. If these could be suitably exploited, Etzler argued, "Any wilderness, even the most hideous and sterile, may be converted into the most fertile and delightful gardens."

Several features of his scheme—including vast projects of terraforming that would leave land masses cut by canals and the oceans covered by floating islands—vividly illustrate the Holocene recklessness with the earth. His was a planetary vision for population growth without limit, fueled without labor; he thought the earth could support a thousand times its population, living in all quarters of the globe, including the poles. But this led him to something very similar to the contemporary project of a green grid, since all that growth would be driven by limitlessly renewable power sources. As his biographer notes, "Etzler designed not a world to come, but the world that came," primarily because he saw the future as "the application of technology to convenience and leisure."[2]

A British edition of *The Paradise Within Reach of All Men* appeared in 1836, and after another edition appeared in 1842, *The United States Magazine and Democratic Review* ran a long review essay as its opening

2. Steven Stoll, *The Great Delusion: A Mad Inventor, Death in the Tropics, and the Utopian Origins of Economic Growth* (New York: Hill and Wang, 2008).

item in November 1843. The essay was unsigned, but its author was the young Henry Thoreau, then serving as a tutor to Emerson's nephew on Staten Island. Thoreau had at this point published almost nothing and was trying to break into the New York literary scene. (The review was as far as he got.) He reviewed Etzler's book with mock appreciation ("We confess that we have risen from reading this book with enlarged ideas, and grander conceptions of our duties in this world"), dry sarcasm ("Let us not succumb to nature. . . . We will wash water, and warm fire, and cool ice, and underprop the earth. We will teach birds to fly, and fishes to swim, and ruminants to chew the cud. It is time we had looked into these things"), and, along the way, deep skepticism. Thoreau objects to Etzler's appeal for corporate forms of organization. He darkly imagines the "environs" of large-scale power-generating enterprises. He notes that Etzler's reforms "prescribe for the globe" but have forgotten its other inhabitants ("What is the part of magnanimity to the whale and the beaver? Should we not fear to exchange places with them for a day?"). He sees that Etzler's proposals have no self-limiting principle and thus open the door to planetary catastrophe: "Who knows but by accumulating the power until the end of the present century, using meanwhile only the smallest allowance, reserving all that blows, all that shines, all that ebbs and flows, all that dashes, we may have got such a reserved accumulated power as to run the earth off its track into a new orbit, some summer, and so change the tedious vicissitude of the seasons?" But mostly he worries that by imagining the globe as a source of power to a humanity that it defines only in terms of its appetite for power and comfort, Etzler's utopia of unlimited emergence hollows out its subject. "It would seem," he declares, "that there is a transcendentalism in mechanics as well as in ethics."[3]

3. Henry David Thoreau, "Paradise (to Be) Regained," in *Reform Papers*, ed. Wendell Glick (Princeton: Princeton University Press, 1973), 19–48.

The opposition between mechanics and ethics that Thoreau draws here has been with us ever since. The green grid that is currently the holy grail of energy reform is, funnily enough, pretty nearly the same utopia of unlimited natural energy that Etzler imagined. This is an instructive irony. The whole tradition of environmental thought has been so reanimated by climate change that we are tempted to think it was always already about greenhouse gases. But its ethical orientation has generally been quite other. While most arguments for renewable energy appeal to a consequentialist ethics—I must cut my carbon footprint because greenhouse gases cause harm—we have seen that ethics so imagined will have no objection to life on the grid as long as it can be decarbonized. Thoreau, of course, could not have known about CO_2 emissions and climate change. But confronted with an energy utopia, he began to imagine himself as a "moralist" of a different kind: one whose task would be to counteract the "transcendentalism in mechanics" that he saw in Etzler.[4]

What did Thoreau mean by "a transcendentalism in mechanics"? He was no simple antimodernist, and he was hardly opposed to technology: He made his living partly through the family business of pencil manufacturing and took special interest in the technical engineering parts of the work. He worked as a surveyor and knew a thing or two about abstract space. And he maintained a serious engagement with science, as Laura Walls and others have shown. The problem with "a transcendentalism in mechanics," in other words, was not the mechanics itself, but the transcendentalism in it; he saw that Etzler's energy projects were not just a means, but had the power to redefine their ends. The narrative of progress in Etzler has a species subject at its core, and it frames that species by its mastery of the planet and the

4. The best account of Thoreau's relation to Etzler ("*Walden* was his counterproposal") remains Sherman Paul, *The Shores of America: Thoreau's Inward Exploration* (Urbana: University of Illinois Press, 1958), 151–57.

elements, its exercise of power as a generalized medium, its evacuation of the question of all ends other than capacity and convenience.

What did "ethics" mean to Thoreau, that he saw it as rivaled by utopias of renewable energy? The Etzler review ends with a rather soggy paragraph about love, which suggests he did not yet entirely know that answer, but he did know what he was up against. Etzler's project was the unholy mirror of his own. Two years after the Etzler review came out, Thoreau built his cabin on the northeast corner of Walden Pond.

Walden has not generally been seen as taking part in the emerging science of energy, but its long first chapter ("Economy") is really a sustained meditation on thermodynamics.[5] Taking inspiration from Justus von Liebig's *Animal Chemistry,* Thoreau evaluates the four "necessaries of life"—food, shelter, clothing, fuel—as forms of the conservation of heat. "Man's body is a stove," he writes, summarizing Liebig; the four necessaries are then treated as means of the conservation of heat. "While Food may be regarded as the fuel which keeps the fire within us,—and Fuel serves only to prepare that Food or increase the warmth of our bodies by addition from without,—Shelter and Clothing also serve only to retain the *heat* thus generated and absorbed." "The expression, *animal life,*" he asserts, "is nearly synonymous with the expression, *animal heat.*"[6] Thus *Walden* begins its infrastructural reflection by contemplating the conservation and transformation of what would very soon come to be called energy, in a way that bridges the life-processes of the body with its technological envelope. In contrast with Etzler, Thoreau's aim is not to reorganize nature around human convenience but to place the basic endeavors of living within natural processes; not to eliminate labor but to

5. I owe much here to an unpublished paper by Christopher Hoogstraten, "'Man's Body Is a Stove': Reading Thoreau's Economy of Energy in the Anthropocene."

6. Henry D. Thoreau, *Walden, 150th Anniversary Edition,* ed. J. Lyndon Shanley (Princeton: Princeton University Press, 2004), 13.

identify its economic surplus; and to reframe such goals as comfort as part of the conservation rather than waste of energy.

Walden sets a high bar for *awareness* as an ethical end. Hence the organizing metaphors of waking and morning. "Every man is tasked to make his life, even in its details, worthy of the contemplation of his most elevated and critical hour" (90). This is clearly an impossible state, but that is part of the point: It is a task. What immediately precedes the line makes this clear: "It is something to be able to paint a particular picture, or to carve a statue, and so to make a few objects beautiful; but it is far more glorious to carve and paint the very atmosphere and medium through which we look, which morally we can do. To affect the quality of the day, that is the highest of arts" (90).

His attempt to re-embed moral consciousness in awareness of planetary conditions is, as Jedediah Britton-Purdy puts it, a version of the Romantic politics of consciousness. "To be awake is to be alive" (90). But what is surprising is the range of experiments in awareness he generates from the polarity of living and dying. Thoreau pursues the meaning of aliveness to encompass attunement to the embodied conditions of perception, attunement to the variety of life (because without this we do not really know our own life), the priority of engagement with the elements over comfort, and unflinching recognition of the world in its least hospitable zones. To track these experiments would be to track all of his writing, as others have done. My purpose here is to remind ourselves that environmentalism in Thoreau is fundamentally an environmental *awareness*. (A modern descendant of this theme is Dale Jamieson's argument for "mindfulness.") This priority of awareness in Thoreau's environmental ethics would seem to be the very opposite of the entitled and normal obliviousness I have been tracing in grid subjectivity and more generally in the infrastructure that buffers us, habituates us, supplies us with means without ends, and mediates our relation to other users. What Thoreau's Etzler review allows us to see is that this dialectical opposition was there from the beginning. He saw his project as unfolding the ethical space closed down by the transcendentalism in mechanics.

In its reflection on the necessary conditions and resources of life, "Economy" continually struggles against the conditions of normal unconsciousness. For Thoreau, this means untangling means and ends in modern infrastructure:

> As with our colleges, so with a hundred "modern improvements"; there is an illusion about them; there is not always a positive advance. The devil goes on exacting compound interest to the last for his early share and numerous succeeding investments in them. Our inventions are wont to be pretty toys, which distract our attention from serious things. They are but improved means to an unimproved end, an end which it was already but too easy to arrive at; as railroads lead to Boston or New York. We are in great haste to construct a magnetic telegraph from Maine to Texas; but Maine and Texas, it may be, have nothing important to communicate. (52)

"Improved means to unimproved ends" is essentially a summary of the Etzler review. Thoreau has several concerns here. He sees that the whole enterprise is underwritten by a narrative of inevitability; that infrastructure takes the place of ends; that media elicit users. Thus "men have become the tools of their tools" (37). Seeing in railroads, the telegraph, power systems, and other user infrastructures a generalization of means, he took as his own goal (in Britton-Purdy's apt phrase) "to develop a relentless practice of exploring and transforming one's own purposes."[7]

7. Jedediah Purdy, *After Nature: A Politics for the Anthropocene* (Cambridge, MA: Harvard University Press, 2015), 141.

Perhaps the background of the Etzler review is why Thoreau's investigation into the ethical meaning of natural life begins with infrastructure—the building of the house itself. "Most men appear never to have considered what a house is" (35). Perhaps, too, that is why he chose such a strange site for it: not many feet from the tracks of a railroad line that was then still under construction. Given the power of the myth of Thoreau as a hermit retreating to the woods, it is easy to forget how significant that decision was. Along with the house itself, the railroad stands as a type of infrastructure. *Walden* reminds us again and again of that railroad. Almost every chapter mentions it. Thoreau tells us that he walked to town by way of the tracks. He notes how it organizes space differently from the cart tracks that cross it. He notes how its sounds and its timetables have reorganized his neighbors' awareness of time. He tries to bring into consciousness the dispersed geographies of commerce and long supply lines it connects, on scales up to the limit of the global (carrying Walden ice, he imagines, to the banks of the Ganges). And he reminds us of its materiality, pausing to honor the workmen forced to clear its tracks from snow. The railroad is one of the most conspicuous features of the environment as Thoreau represents it. Even the climactic passage in the "Spring" chapter about the thawing sandbank—emblem of organic vitality—is set in the cut of the railway embankment.

Thoreau treats the railroad in a double aspect. On one hand, he saw its progressive appeal; the encomium to "commerce" as he watches the train go by has been a puzzle to critics ever since Leo Marx's *The Machine and the Garden*. More familiar to environmental thinking is its other face: The railroad stands in for all modern social systems that threaten to defeat basic ethical aspirations scaled to consciousness.

> To make a railroad round the world available to all mankind is equivalent to grading the whole surface of the planet. Men have

> an indistinct notion that if they keep up this activity of joint stocks and spades long enough all will at length ride somewhere, in next to no time, and for nothing; but though a crowd rushes to the depot, and the conductor shouts "All aboard!" when the smoke is blown away and the vapor condensed, it will be perceived that a few are riding, but the rest are run over—and it will be called, and will be, "A melancholy accident." (53)

Here again Thoreau notes how we are drawn into unconsciousness by the mysterious inevitabilism of a medium that is always about to deliver a future. As he puts it when he returns to the railroad in "Sounds," "We have constructed a fate, an *Atropos,* that never turns aside" (118).

Thoreau was acutely aware that his project of bringing the elemental/perceptual conditions of existence into awareness was already countercultural. And indeed his central preoccupation with the de-ethicalizing effects of infrastructure was mostly lost in the long tradition of environmental thought that followed in the wake of *Walden.* The major exception to that last claim is the countercultural movement of the 1960s and 1970s to which we owe the resurgence of *Walden* as a foundational text.

FROM *WALDEN* TO THE OFFGRID MOVEMENT

Each extension of the grid gave rise to resistance, and each new layer of the grid gave a new meaning to being off it. Wolfgang Schivelbusch notes a widespread resistance to gas lighting in the main rooms of nineteenth-century European bourgeois homes (a feeling shared by Edgar Allan Poe in "The Philosophy of Furniture"). People preferred the paraffin lamp apparently for no better reason than that "by keeping their independent lights, people symbolically distanced themselves from a centralized supply." Similarly with real estate: The upper part of the Manhattan grid, now enclosing Central Park, was not yet built when Frederick Law Olmsted designed the park as a negation of what would soon surround it. "The time will come when New York will be built up," he wrote at the time, "when all the grading and filling will be done, and when the picturesquely-varied, rocky formations of the Island will have been converted into formations for rows of monotonous straight streets, and piles of erect, angular buildings."[8] From that time to the present, grid architecture has been one of the poles against which environmental consciousness attempts to free itself, often through a language of localism and a sense of "place" rather than "space."

Given the invisibility of the electrical power grid, it is perhaps unsurprising that in the long century of grid modernity, between Thoreau and the emergence of climate change, very few major thinkers noticed it—even in fields like media theory, political theory, and environmental philosophy, where it might have been central. Two exceptions stand out: Lewis Mumford and Martin Heidegger. Mumford, in 1934, appealed to the electric grid as the main reason to think that modernity had entered a qualitatively different technical phase—in his idiom, "neotechnics" rather than the "paleotechnics" of the industrial era. "The neotechnic phase was marked to begin with, by the conquest of a new form of energy: electricity."[9] And for Mumford, there was a direct connection with media: "Instantaneous personal communication over long distances is one of the

8. Frederick Law Olmsted, "To the Board of Commissioners of the Central Park" (Letter of May 31, 1858), in *The Papers of Frederick Law Olmsted*, vol. 3, *Creating Central Park, 1857–1861*, ed. Charles E. Beveridge and David Schuyler (Baltimore: Johns Hopkins University Press, 1983), 196.

9. Mumford, *Technics and Civilization*, 221.

outstanding marks of the neotechnic phase."[10] This insight, however, got lost in much subsequent thought about technology.

Like Mumford, Heidegger in "The Question Concerning Technology" (1953) argues that the grid separates modern technology from the technological stances of the past. As almost all commentators note, Heidegger's central example is the hydroelectric dam on the Rhine. What fewer notice is that the dam is not just any example of technology; Heidegger singles it out because what it represents is a constancy of available energy—that is, the grid.

> The hydroelectric plant is set into the current of the Rhine. It sets the Rhine to supplying its hydraulic pressure, which then sets the turbines turning. This turning sets those machines in motion whose thrust sets going the electric current for which the long-distance power station and its network of cables are set up to dispatch electricity. In the context of the interlocking processes pertaining to the orderly disposition of electrical energy, even the Rhine itself appears as something at our command.[11]

The last clause here seems to imply a familiar indictment of the human dominion over nature as "something at our command," an

10. Mumford, *Technics and Civilization,* 241. Just before this he writes: "What will be the outcome? Obviously, a widened range of intercourse: more numerous contacts: more numerous demands on attention and time. But unfortunately, the possibility of this type of immediate intercourse on a worldwide basis does not necessarily mean a less trivial or a less parochial personality. For over against the convenience of instantaneous communication is the fact that the great economical abstractions of writing, reading, and drawing, the media of reflective thought and deliberate action, will be weakened" (240; this is extended, on 241, to a worry about the fascistic possibilities, in radio, of a relation between ruler and ruled that seems immediate and direct but is in fact secondary and partial). Mumford also argues that the neotechnic phase introduces greater concern for conservation and ecology (255–59). But he thought that one sign of this, ironically, was "the building up in agriculture of an appropriate artificial environment" (258).

11. Martin Heidegger, *The Question Concerning Technology and Other Essays,* trans. William Lovitt (New York: Harper, 1977), 16. Further citations to this text will be parenthetical.

arrogance of the will; and that is how the passage is sometimes interpreted. But the point of the sentence is that this dominion takes a different form *in the context of* the "orderly disposition of electrical energy." "At our command" carries more of the sense of being "on call"; Heidegger is leading up to the notion of the "standing reserve." "Everywhere everything is ordered to stand by, to be immediately at hand, indeed to stand there just so that it may be on call for a further ordering" (17).

Three paragraphs before Heidegger gets to the hydroelectric dam, he has already made clear that he has in view this difference between the dam and earlier power technologies. He ventriloquizes the objection, "But does this not hold true for the old windmill as well?" in order to clarify the distinction: "No. Its sails do indeed turn in the wind; they are left entirely to the wind's blowing. But the windmill does not unlock energy from the air currents in order to store it" (14). Of course, the windmills of today do exactly that; they are now part of the grid. The point is that storage, by making energy constantly available, changes our relation to wind and water. Even "the coal that has been hauled out in some mining district has not been supplied in order that it may simply be present somewhere or other. It is stockpiled; that is, it is on call, ready to deliver the sun's warmth that is stored in it" (15). This "fundamental characteristic" (14) of a standing reserve is different from those modes of manipulation in which human subjects stand over nature as an object. Human dominion is a matter of "enframing" rather than will, or tool-using, or even the machine. "Whatever stands by in the sense of standing-reserve no longer stands over against us as object" (17); it "disappears into the objectlessness of standing-reserve" (19).

Heidegger's essay has a complicated understanding of history; even though "exact science" arose two centuries before the technology of energy, he argues, "Modern technology, which for chronological reckoning is the later, is, from the point of view of the essence holding sway within it, the historically earlier." One reason for this counterintuitive storyline is that the enframing that Heidegger sees in the grid is not simply the expression of rationalism or the will to dominate, as many environmental ethicists have thought. "Modern physics is the herald of Enframing, a herald whose origin is still unknown. The essence of modern technology has for a long time been concealing itself, even where power machinery has been invented, where electrical technology is in full swing, and where atomic technology is well under way" (22). Enframing, though manifest in the infrastructure of power, both came before it and is yet to come.

Given the terms of Heidegger's critique, the windmills of today, like solar panels, would have to be seen not as a break from the grid of the hydroelectric dam but as more of the same. (Indeed, despite all the environmental arguments against hydroelectric dams, they are now counted as renewable energy. As with nuclear power, this is a topic of fierce contention among environmentalists.) So one might be tempted to conclude that his reflection on technology and infrastructure does little to clarify the basic dilemma of greenhouse gas emissions, which he could hardly have foreseen. Like Thoreau, however, Heidegger was thinking about the ways our infrastructure involves us in an enframing of being that has something like "man" as its subject, and involves us in the form of our consciousness to such a degree that it is difficult to imagine wresting free from it.

Heidegger also writes as though the enframing that power represents descends upon "man" in general, globally; the offgrid billion make no appearance in his thought. Yet he might have been enabled in this profound perception of the condition of the grid by the fact that while writing "The Question Concerning Technology" he was shuttling between Freiburg, on the grid, and his second house (often referred to as a "hut," but really a house)—an offgrid structure at Todtnauberg. This house was apparently connected to public utilities only later in Heidegger's career. He used a hand pump for water, an earth closet in lieu of a sewer, and lamps for light. His ability to see the strangeness of grid modernity, in other words, may well owe its clarity to this second, offgrid location.

The term "offgrid" (along with its variants) first came into circulation in the era of the Tennessee Valley Authority. It was the negative shadow of the grid: to describe an area as off the grid was to name it as a target for grid expansion. The term still carries this charge in developing countries engaged in rural electrification (though the introduction of microgrids, like community solar and other developments in infrastructure design, softens the on-/offgrid distinction). Only in the 1960s, and especially in the United States, did "offgrid" come to have positive associations with an ethical and political vision that would motivate distributed and renewable power, scaled to use.

To the extent that it survives in US culture, this idealized state of being offgrid can mean a lot of things. To many, it means getting away from the internet. *Offgrid* is the title of a survivalist magazine, preparing its prudent subscribers for the anarchic remnant in the last of days. Promoting the offgrid life can also be a branding association for celebrities, the benevolent marketing of a romantically public fantasy of privacy. In each case a different perception of the grid motivates the desire to be off it—though in the age of GPS it is less clear what *is* off it.[12]

For the second half of the twentieth century, photovoltaic solar power and wind turbines grew in the various shadows of the grid, especially in the United States. Partly this was by default, because fossil fuel interests so completely dominated policy and public media. But their development—and especially their domestication—was also motivated in the United States, to a far greater degree than elsewhere in the developed world, through the critical reaction against

12. On GPS, its reorganization of space, and the form of utilities, see William Rankin, *After the Map: Cartography, Navigation, and the Transformation of Territory in the Twentieth Century* (Chicago: University of Chicago Press, 2016).

the expansion of power consumption and its corporate/utility administration, a reaction that was central to the emerging environmental movement. Offgrid spaces came to be sought out as the laboratory not just for environmental consciousness but for alternative infrastructures motivated by environmental concern. This legacy therefore represents the most significant context for thinking about the social and planetary meaning of the grid, even though (or especially because) that legacy is now thrown in question by the very integration of renewables on the grid that has become critical to climate change mitigation.

Here too, deeper histories are in play. Solar power in general has of course a history longer than history itself. Solar generation of transmissible power dates from industrialization and grew in the footsteps of fossil fuels. Many of the early versions, such as the one imagined by Etzler or the one operated by August Mouchot in 1865, worked by concentrating heat through reflectors to generate steam. Sometimes the lure for such schemes was free power, but most of the early inventors cited anxiety about fossil fuel shortages. In 1860, before building his solar motor, Mouchot wrote, "Eventually industry will no longer find in Europe the resources to satisfy its prodigious expansion. Coal will undoubtedly be used up. What will industry do then?"[13] John Ericsson (inventor of the ironclad *Monitor*) similarly experimented heavily with solar power for steam because he foresaw an unsound global dependence on coal.[14] (That was just before oil wells came along.)

Photovoltaic solar came into possibility, interestingly, in the same years that Edison was working on electricity; and it, too, had

13. Quoted in Charles Smith, "Revisiting Solar Power's Past," *Technology Review* 98, no. 5 (July 1995): 40.

14. John Perlin, *From Space to Earth: The Story of Solar Electricity* (Cambridge, MA: Harvard University Press, 2002), 5.

a connection to the telegraph. The chief electrician laying the transatlantic cable in the 1860s, Willoughby Smith, created "a superior device for detecting flaws in the cables as they were submerged," made with crystalline selenium. In underwater testing the device worked fine—until the sun came out. Smith eventually worked out that the selenium was generating electric charge in response to light. Nothing much came of it until Edison started building his grid; then, a New Yorker named Charles Fritts created the first photovoltaic cell, using thinly sliced selenium. As he made clear in a triumphal press announcement of 1885, he envisioned it as a rival to Edison's grid. "We may ere long see the photoelectric plate competing with" coal-fired electric plants, he said.[15] Curiously, though, he did not envision his solar modules as a rival grid. Solar arrays, he wrote, "are intended principally for what is known as 'isolated' working, i.e., for each building to have its own plant."[16] Solar power, from the moment of its inception, was to be electricity off the grid.

And there things sat during the long heyday of grid construction. Important technical advances in PV cells were made in the 1940s and 1950s. The first practical applications were in the most literally offgrid spaces: outer space and offshore. NASA funded PV cell development when it started launching satellites, realizing that batteries would turn the entire satellite into waste when they ran out—as the first two Sputniks did, about a week after they went into orbit. The other principal site of installation was offshore oil rigs, which were mandated by the Coast Guard to maintain signal lights; so oil companies ironically enough became the first major purchasers of PV technology after NASA.

It was only in the postwar decades that PV electric power made its way into domestic space. Just as satellites and offshore rigs provided

15. Perlin, *From Space to Earth*, 17.
16. Perlin, *From Space to Earth*, 75.

space for industrial solar, domestic PV solar for several decades developed almost exclusively in offgrid sites. Offgrid life had become newly motivated. In the same period in which Americans were becoming accustomed to higher and higher consumption of power, a critical countermovement developed around the linkage between state and corporate administration, consumption and the environment. Lewis Mumford's career illustrates the connection. He began life wanting to be an electrical engineer, in the heyday of regional planning; in his 1934 *Technics and Civilization*, he envisioned a future republic led by engineers. But he came to reassess his optimism, partly because in the same years, from 1920 to 1970, electricity consumption was doubling every decade.[17] By the time of *The Myth of the Machine* (1970), Mumford was denouncing complex technical systems of all kinds as destructive and amoral realizations of a pathological desire to dominate nature and people alike.[18]

By that point, people had begun to occupy offgrid spaces with intent. It is difficult to generalize about the so-called counterculture of offgrid experiments in these years, since it ran a gamut from pastoralist utopians to the technophiles who created cyberculture.[19] A wide range of intellectual sources circulated across these strata, from Aldo Leopold to Buckminster Fuller. Most of these thinkers barely noticed the grid; or rather they responded to it only indirectly.[20] Despite the differences among them with respect to technology in general, solar electricity appealed to many on this spectrum, and technology transfer among these variously motivated people was

17. Nye, *Electrifying America*, 198.

18. In the years between these two books, similar perceptions of grid modernity were articulated by Herbert Marcuse (*One-Dimensional Man*, 1964), Jacques Ellul (*The Technological Society*, 1966), Theodore Roszak (*The Making of a Counter Culture: Reflections on the Technocratic Society and Its Youthful Opposition*, 1969), and Charles Reich (*The Greening of America*, 1970).

19. See Turner, *From Counterculture to Cyberculture*.

20. E.g., Marcuse in *One-Dimensional Man*.

the aim of the *Whole Earth Catalog* and the extensive social networks around it. Offgrid solar was to some an alternative to fossil fuels, especially after the OPEC crisis of the mid-1970s; but to others it was more: an alternative to the way the power grid was oriented to maximize consumption, the way it entailed an administered society, and the way it organized abstract space. In the utopian left vector by which solar technology moved from satellites and oil rigs to domestic power, people were motivated to make sacrifices in both cost and efficiency in order to build infrastructure that implied a different social future. They did so in a language of ethical environmentalism, speaking of localism, stewardship, and awareness in opposition to the abstract space, privatization, consumerism, and unconsciousness of the power economy; and they saw offgrid spaces as places that could be occupied by such virtues only if they could be organized by small-scale and sustainable infrastructure.

Although it is now mostly forgotten, the best-elaborated version of this vision was a movement broadly labeled "appropriate technology"—perhaps the most serious engagement between environmentalism and infrastructure.[21] Its intellectual sources ranged from

21. The best secondary treatment for the United States is Andrew G. Kirk, *Counterculture Green: The "Whole Earth Catalog" and American Environmentalism* (Lawrence: University Press of Kansas, 2007). See also Frank Laird, "Constructing the Future: Advocating Energy Technologies in the Cold War," *Technology & Culture* 44, no. 1 (Jan. 2003): 27–49; and Leigh Glover, "From Love-Ins to Logos: Charting the Demise of Renewable Energy as a Social Movement," in *Transforming Power: Energy, Environment, and Society in Conflict*, ed. John Byrne, Noah Tolly, and Leigh Glover (New Brunswick: Transaction Publishers, 2006), 249–70. Primary sources include Richard S. Eckaus, *Appropriate Technologies for Developing Countries* (Washington, DC: National Academy of Sciences, 1977); Peter Harper, Godfrey Boyle, and the editors of Undercurrents, eds., *Radical Technology* (New York: Pantheon, 1976); David Dickson, *Alternative Technology and the Politics of Technical Change* (London: Fontana, 1974); Ken Darrow and Rick Pam, *Appropriate Technology Sourcebook* (Stanford, CA: Volunteers in Asia, 1976); Steve Baer, *Sunspots* (Albuquerque: Zomeworks, 1975); Hugh Nash, ed., *The Energy Controversy: Soft Path Questions and Answers* (San Francisco: Friends of the Earth, 1979); J. Baldwin and Stewart Brand, eds., *Soft-Tech* (New York: Penguin, 1978). The appropriate-technology movement inspired its own internal critiques in Langdon Winner, *The Whale and the Reactor: A*

Thoreau and Gandhi to Leopold and E. F. Schumacher, and encompassed thinkers as different as Amory Lovins and the anarchist utopian Murray Bookchin.[22] The organizing slogan of "appropriate technology" grew out of the international development controversies of the 1960s, in which context it was first known as "Intermediate Technology." Its early formulations had to do with the need for less capital-intensive and high-maintenance technologies for the developing world—that is, for sites "intermediate" between the developed nations and the rural poor of undeveloped countries. It nourished practical experiments in infrastructure both in the United States and in the developing world, often with an explicit coupling. The effort was to develop techniques that would work in both locations—the offgrid sites in wealthy nations of the North and the necessarily offgrid sites in poorer ones of the South—with ways of living scaled accordingly.[23] As the historian Andrew Kirk writes, "For supporters of appropriate technology, the most radical action one could take against the status quo was not throwing bombs or staging sit-ins, but fabricating wind generators to 'unplug from the grid.'"[24]

Even at the heyday of appropriate technology, however, the road forked. The best way to see this is to compare Schumacher's *Small Is Beautiful* (1973) with Amory Lovins's *Soft Energy Paths,* which he later followed up with a deliberate echo of Schumacher in *Small Is Profitable* (2002). Schumacher worked for the British National

Search for Limits in an Age of High Technology (Chicago: University of Chicago Press, 1986), especially "Building the Better Mousetrap," on appropriate technology; Landon Winner, *Autonomous Technology: Technics-out-of-Control as a Theme in Political Thought* (Cambridge, MA: MIT Press, 1978); Witold Rybczynski, "Appropriate Technology: The Upper Case Against," *CoEvolution Quarterly* 13 (Spring 1977): 81–83; and Witold Rybczynski, *Paper Heroes: A Review of Appropriate Technology* (Garden City, NY: Anchor, 1980).

22. E. F. Schumacher, *Small Is Beautiful: Economics as if People Mattered* (New York: Harper, 1973); Murray Bookchin [pseud. "Lewis Herber"], *Our Synthetic Environment* (New York: Knopf, 1962); Murray Bookchin, *Post-Scarcity Anarchism* (San Francisco: Ramparts, 1971).

23. Witold Rybczynski traces some of these connections in *Paper Heroes.*

24. Kirk, *Counterculture Green,* 31.

Coal Board from 1950 to 1970 while also serving as a UN economic adviser to Burma and India; the misfit between their infrastructural needs and British models of large-scale power led him to found the Intermediate Technology Development Group in 1963. He claimed derivation from Gandhi, as did much of the appropriate-technology movement. (Although connections were forged with some Gandhian associations, the movement flourished most in developed nations and in California especially.) From Gandhian as well as Buddhist sources, he formed an idea of "nonviolent technology" as an alternative to centrally organized, state-administered, large-scale systems. "As the world's resources of non-renewable fuels—coal, oil and natural gas—are exceedingly unevenly distributed over the globe and undoubtedly limited in quantity, it is clear that their exploitation at an ever-increasing rate is an act of violence against nature which must almost inevitably lead to violence between men."[25] Nonviolent energy infrastructure would need to be small-scale, distributed, built to existing ways of life, repairable and maintainable by users, and designed to facilitate reverence toward nonhuman life.

Lovins first came to fame by a 1976 essay in *Foreign Affairs* titled "Energy Strategy: The Road Not Taken?," the essential themes of which he has since elaborated in a long string of articles and books, including the influential *Reinventing Fire* of 2011. "Soft energy" for Lovins means infrastructures that (1) rely on renewable energy resources, (2) are diverse and designed for maximum effectiveness in particular circumstances, (3) are flexible and relatively simple to understand, (4) are matched to end-use needs in terms of scale, and (5) are matched to end-use needs in terms of quality. Thus he argued for distributed generation in terms that seem very similar to Schumacher's, and in a number of works (such as *Non-Nuclear Futures*, in 1975) the argument was explicitly cast as one of "ethical

25. Schumacher, *Small Is Beautiful*, 64–65.

concern." But Lovins's strategy was to persuade policy elites, and his way of doing so was to argue for the greater efficiency and rationality of soft energy. He became the prophet of the "smart" grid:

> Efficiency is about smarter technologies that do more with less—not privation, discomfort, or curtailment. Demand response voluntarily alters customers' consumption patterns in response to the changing price of electricity over time, incentive payments, or other signals of scarcity or abundance, saving money for both customers and provider. In our modeling, we have assumed only demand response methods that are *unobtrusive, so the customer doesn't mind or notice.*[26]

Smart grid, dumb users.

Langdon Winner observes that the economic appeal of the argument silently replaces the ethical choices Lovins had tried to pose:

> [Lovins] notes that basing energy decisions on social criteria may appear to involve a "heroic decision," that is, "doing something the more expensive way because it is desirable on other more important grounds than internal cost." But Lovins is careful not to appeal to his readers' sense of courage or altruism. "Surprisingly," he writes, "a heroic decision does not seem to be necessary in this case, because the energy system that seems socially more attractive is also cheaper and easier." But what if the analysis had shown the contrary? Would Lovins have been prepared to give up the social advantages believed to exist along the soft energy path? Would he have accepted "centrism, vulnerability, technocracy, repression, alienation," and the like? Here Lovins yielded ground that in recent history has again and again

26. Lovins, *Reinventing Fire,* 170, my emphasis.

been abandoned as lost territory. It raises the question whether even the best intentioned, best qualified analysts in technological decision making are anything more than mere efficiency worshippers.[27]

The subtle but all-important shift of emphasis between Schumacher and Lovins indicates the fate of offgrid appropriate technology: It is now on the grid. The connection between renewable energy and the grid is a relatively new phenomenon, one that

27. Winner, *The Whale and the Reactor*, 53–54. Writing in 1983, Winner imagined the development of solar power along a number of different paths and accurately foresaw that this fateful choice would never be offered to the public. "We can expect to see events unfold in our lifetimes with outcomes that could have many different dimensions. If solar cells become feasible to mass produce, if their price in installed systems comes down to a reasonable level, solar electricity could make a contribution to our society's aggregate energy needs. If the day arrives that photovoltaic systems are technically and economically feasible (and many who work with solar electric prototypes believe they will be), there will be—at least in principle—a choice about how society will structure these systems. One could for example, build centralized photovoltaic farms that hook directly into the existing electrical grid like any other form of centrally generated electric power. It might also be possible to produce a great number of stand-alone systems placed on the rooftops of homes, schools, factories, and the like. Or one could design and build medium-sized ensembles, perhaps at a neighborhood level. When it comes time to choose which model of photovoltaic development our society will have, a number of implicit questions will somehow be answered. How large should such systems be? How many will be built? Who should own them? How should they be managed? Should they be fully automatic? Or should the producer/consumer of solar power be actively involved in activities of load management?" ("*Technē* and *Politeia*: The Technical Constitution of Society," in *Philosophy and Technology*, ed. Paul T. Durbin and Friedrich Rapp [Dordrecht: D. Reidel, 1983], 108). "Here is an opportunity to extend responsibility and control to a greater number of people, an opportunity to create diversity rather than uniformity in our sociotechnical constitution. . . . Rather than pursue the lemming-like course of choosing only that system design which provides the least expensive kilowatt, perhaps we ought to consider which system might play the more positive role in the technical infrastructure of freedom. It goes without saying that the agencies now actually developing photovoltaics have no such questions in mind. . . . As we have done so often in the past, our society has, in effect, delegated decision-making power to those whose plans are narrowly self-interested. One can predict, therefore, that when photovoltaic systems are introduced they will carry the same qualities of institutional and physical centralization that characterize so many modern technologies" (*The Whale and the Reactor*, 57).

threatens to eliminate offgrid spaces entirely. The infrastructural turning points were net metering and grid-tied solar, which were originally progressive initiatives to promote renewables. (The earliest movement in this direction was undertaken by the state of Minnesota, in 1983.) By the early twenty-first century, this policy had spread so far as to uncouple alternative energy from its countercultural politics, at the very moment when greenhouse gases became the paramount environmental concern. The resulting shift in the politics of the grid highlights the essential discontinuity or even contradiction between twentieth-century environmental thought and twenty-first century climate change. As recently as 2004 Hermann Scheer could write, in *The Solar Economy*, that solar energy has the promise of preserving the human gains of energy-intensive modernity without a grid. "Only a solar global economy can satisfy the material needs of all mankind and grant us the freedom to re-establish our social and democratic ideals" (32).[28] Unfortunately for Scheer and others who imagined that climate change would invigorate opposition to the technical modernity of the grid, the growth of renewable energy seems to require greater integration of the grid, not less. And the role once given to ethics now falls to efficiency.

There are several reasons for this. Wind power has come to be dominated by highly capitalized, massive installations. Both wind and solar are highly variable, which means among other things that peak production does not match peak demand. They therefore need to be supplemented with what is called "dispatchable power," available at all times both to meet demand peaks and to maintain a synchronous grid. Because they vary with weather at the point of generation, they are less predictable and more locally sensitive. They carry all the promise of distributed production, but to use distributed

28. Hermann Scheer, *The Solar Economy: Renewable Energy for a Sustainable Future* (London: Earthscan, 2004), 32.

production as a means for decarbonizing the grid requires efficient coordination. As a result, many of the changes now in motion among utilities, regulators, and advocates have a common goal that has come to be called "the smart grid." This name gets applied to a variety of innovations: microgrids embedded within the grid, larger spheres of interconnection among grids, demand management and "smart meters," and so forth. Localism at either end of the circuit (generation, distribution) leads to greater complexity and regional integration in the middle (transmission, or the grid).[29] Again, Bakke puts it well: "Distributed solar causes the once 'somewhere' of electricity production to start to look a lot like the more familiar 'everywhere' of electricity consumption."[30]

It is ironic, then, that the offgrid movement—perhaps the most significant experiment in sustainable ethics and infrastructure yet attempted in developed nations—went into eclipse at just the moment when the realization of global warming might have given it most urgency. Doubly ironic, because being offgrid has never been easier. Indeed, some power companies have begun to worry publicly about the economics of "grid defection." The vision of a solar revolution was for many years kept alive by an offgrid movement that is now obscured at the moment of its own technical realization. Many of its technical innovations survive, but the offgrid spaces in which the vision was made practical are themselves increasingly on the grid, while its ethico-technical challenges are remembered, if at all, as fringe pastoralism.

Where then do we stand? Some might think I have merely been sounding the elegiac note, and there is some truth in that; I wish to register a protest against the dismissal of appropriate technology as a

29. The best treatment of this subject is Peter Fox-Penner, *Smart Power: Climate Change, the Smart Grid, and the Future of Electric Utilities* (Washington, DC: Island Press, 2010).
30. Bakke, *The Grid*, 231.

hippie fantasy. Others might see this as an argument for grid defection. Again, fine: Unplug (if you can); it will do you good. Still others might object that since we are nowhere close to realizing the utopia of the green grid, in the meantime the ethical reasons for reducing our power use through awareness continue to hold force. And I agree with this, too; I have undertaken this inquiry in part to understand why so many people leave the power running unthinkingly, despite their avowed concern for carbon emissions.

But I also think that there is a paradox here that needs to be brought into focus for any project of environmental ethics. In the absence of a politics for alternative infrastructure, if we look down the road that is being paved for us with green intentions, we see that we are preparing ourselves for an ever more enveloping grid modernity in which ethicalized awareness will have ever less scope.

COMMENTARIES

Escaping Gridlock

DALE JAMIESON

As I write these comments my contribution to climate change varies greatly depending on when, where, and with what I am writing: sunny midday in San Diego on my solar-powered laptop, 3:00 p.m. Wednesday in New York at my workstation, or marking up the text this morning with a number 2 pencil in a coffeehouse in Berkeley. Yet, wherever and whenever I am writing, my consciousness is dominated by syntax and semantics: Thoughts about my carbon footprint are absent, unless they figure in the content of what I'm writing. For this, we can thank our cognitive systems. Attention is the scarcest of our resources, and it takes drama, expectation, or intention for us to allocate it. We can also thank the electrical infrastructure—the grid—for our cluelessness. For it provides a service whose source is beyond notice and beneath attention.

Warner's central thesis is that the grid has transformed renewable energy's potential for empowerment into even greater disempowerment. While Leopold talked about ethics, the contemporary environmental movement talks about efficiency. Like their colleagues in government, professional environmentalists are more likely to focus on the economic value of ecosystem services than on ecosystems

On the Grid. Michael Warner, Oxford University Press. © Regents of the University of California 2025.
DOI: 10.1093/oso/9780197696248.003.0005

themselves, with their diverse, often nonquantifiable, and sometimes transformational value. The sense of place, celebrated by Thoreau and writers such as Wendell Berry, is increasingly effaced by a placeless global outlook, indexed to constructs such as "mean global surface temperature." As Warner tells us, his lectures are elegiac, an argument for personal grid defection, an exercise in understanding our own dissociated behavior, and most of all a plea for "a politics for alternative infrastructure" that will address the following "paradox . . . for any project of environmental ethics": that "the road . . . being paved for us with green intentions" actually leads to "an ever more enveloping grid modernity in which ethicalized awareness will have ever less scope" (p. 76).

The story that Warner tells is an instance of a more general story about the stealth infrastructure, characteristic of modernity, that has substituted general, impersonal, anonymous, and structurally determined relationships for more particular, personal, face-to-face, episodic interactions. Personal, episodic interactions require high levels of attention, monitoring, and intervention. Impersonal, structurally determined relationships require relatively little attention and infrequent conscious intervention. Like children who are supposed to be seen but not heard, services in the modern world are supposed to be enjoyed, but the infrastructure on which they depend is not supposed to intrude into our consciousness.

Other examples that conform to this general model of substitution include shopping on Amazon versus buying at a street market; using credit cards rather than individual letters of credit; state determinations of guilt or innocence, and sanctioning, versus private systems of justice and conflict resolution; conveying information on social media rather than privately through the post office or even the internet. At the most pedestrian level, even scanners replacing cashiers, and ATMs substituting for bank tellers, are instances of this transformation.

Having made this distinction, I now want to soften it. Street markets, letters of credit, and private systems of justice rely on infrastructure too. In fact, we might say that it's infrastructure all the way down. Still, there are important differences at least in degree and perhaps in kind in the infrastructure that is characteristic of modernity and that which came before.

The shift to impersonal, anonymous structures from personal, face-to-face interaction has produced enormous benefits. Because these impersonal structures typically require less attention, they are often experienced as providing freedom and even a sense of liberation. Since these structures are generally more efficient, they allow us to get more of what we want in goods and services. By helping to create and sustain systems of cooperation among large numbers of strangers, they can have positive spillover effects.[1] These institutions of modernity even sometimes enable us to do things that we otherwise could not have done. Because of the mediation of communication infrastructure, we are now able to vividly see and talk to people on the other side of the world in a way that was impossible a half century ago.

The electrical grid, the primary subject of Warner's lectures, is not just an example of what I am talking about but increasingly the "ur-infrastructure" that supports other infrastructure. From the beginning, the grid was fed by fossil fuels, and its appetite for them has continued to be voracious.[2] Fundamental though it is, the grid hardly ever comes up in small talk or political campaigns. However, its progeny (e.g., climate change) are ubiquitous. Climate change is discussed in national security councils and also feeds and provokes our

1. Paul Seabright, *The Company of Strangers: A Natural History of Economic Life* (Princeton: Princeton University Press, 2004).
2. While the fraction of electricity generated by fossil fuels is slowly declining, the total is not (https://www.eia.gov/todayinenergy/detail.php?id=48896).

anxieties about shopping choices and even about whether to have children. Climate change poses many problems, and that in itself may be the greatest problem of all. Among this plentitude it is the energy problem that is often taken to be most salient. When climate change is seen in this way, the most fundamental challenge is to decarbonize the energy system. From this perspective the grid is at the center, and so also at the center of the gridlock that characterizes our politics. Despite being an unseen character, the grid is a central actor in "the age of climate change."

Climate change is now the dominant theme of environmental politics. This is surprising because in the 1980s and early 1990s the environmental community resisted when atmospheric scientists forced climate change to the center of the environmental agenda. Nuclear power, then almost universally reviled by environmentalists, loomed as one possible response to climate change. Environmental groups also feared that they could not make climate change immediate and visceral enough to fundraise around. Complicating matters further, carbon dioxide—with its global mixing and indirect effects over large and disjointed spatial and temporal dimensions—does not behave like most other pollutants. Furthermore, as Warner hints, even if we were to solve the energy problem, we might still make the world worse by the lights of many environmentalists. In a carbon-constrained world we could still be poisoning ourselves with chemicals and polluting the air and water. Unlimited clean energy could lead to even more thoughtless consumption, waste, and extinction—not to mention sprawl and overpopulation.

We may yet arrive at this future. While many people see climate disruption as the central problem of the Anthropocene, others see climate control (geoengineering) as its central opportunity. From their point of view, the problem is not that humans dominate nature but that we do so clumsily, stupidly, and inefficiently. With good science, smart technology, and a supportive government, help could be on the way. If left to the technocrats, the climate system itself could become another part of our infrastructure.[3]

As Warner points out, for 1960s hippie visionaries the promise of solar energy was that it would enable us to live off the grid and practice something like a Rousseauian way of life. There is nothing intrinsic to the concept of real estate that requires a property to be connected to the grid. When you buy a house, it comes with ceilings, floors, walls, fixtures, and probably appliances. It could also come with an energy system. A good energy system, one that is reliable and doesn't regularly brown out, would add to the value of the house in just the way that central air conditioning, marble countertops, and Japanese toilets add to a house's value. While it is impolite to say it in this way, one of the great successes of the environmental lobby over the last few decades has been to squash the outlaw hippie vision and to transform solar power into feedstock for the grid. Stewart Brand is a living metaphor for this transformation. In 1968 he conceived the *Whole Earth Catalog* as providing "access to tools" for the back-to-the-land movement. In 2009 he wrote a book with the bracing title *Whole Earth Discipline,* in which he advocated for geoengineering, nuclear power, urbanization, and restoring extinct animals by genetic manipulation.[4]

In addition to their benefits, the institutions of modernity that I have been discussing also have costs. They make it difficult to have a sense of personal relationship with those who in some cases provide quite intimate services (e.g., food and clothing). Moreover, just as impersonal, anonymous structures can more efficiently produce benefits, they can also more efficiently produce devastating harms.

3. David Keith, *A Case for Climate Engineering* (Cambridge, MA: MIT Press, 2013). My cautions were expressed early in "Ethics and Intentional Climate Change," *Climatic Change* 33, no. 3 (July 1996): 323–36.

4. See https://en.wikipedia.org/wiki/Whole_Earth_Discipline for information on this volume.

The industrial scale at which humans have slaughtered other humans in wars and genocides and nonhumans for food would not have been possible without the infrastructure of modernity. By buffering individual actions from outcomes and aggregating them, these impersonal structures make it difficult for those involved to feel a sense of responsibility for the systemic consequences of the outcomes to which they contribute. This is part of the logic of climate change, but also of child labor and modern meat production. The fact that there is no emotionally salient or epistemologically vivid nexus between the actions of individual agents and such unwanted outcomes supports us in our denial, and so we are shocked—shocked!—when we are finally brought to understand what is produced by the systems of which we are a part. Even then, participation in these systems can feel coercive since it is so difficult to opt out of them. And so we have the liturgy of the modern age: "What else could I do?" "It doesn't matter what I do." Instead of encouraging a sense of responsibility, these institutions of modernity often create episodes of guilt against a broad background of a sense of complicity that is not systematically connected to action. And so together we destroy the world but no one feels responsible.

The electrical grid, the focus of Warner's lectures, produces enormous benefits that could not have been produced otherwise, and also produces harms and risks, some of which (e.g., climate change) would have been difficult to anticipate. Still, as Warner points out, some of the harms and risks of the grid were clear even before climate change became salient. The example of the first electrified house in New York City, the J. P. Morgan house, already contains the seeds of environmental injustice, with the services enjoyed in the house being produced by the "steam-powered generator [that] sat next to the house, belching smoke from its coal furnace" (p. 37n.38). This is not so different in principle from the Four Corners Power Plant, fed by the Navajo Coal Mine, built in Fruitland, New Mexico, in the 1960s on land owned by the Navajo Nation, producing electricity that goes mainly to Arizona and California. The plant produces a great deal of local pollution (e.g., NOx and SOx) that primarily affects the population living near the plant. Yet tribal leaders have opposed the rapid shutdown of the plants because of the economic benefits they produce for the tribe and local communities.

The cost of the grid that most interests Warner is its role in diminishing "ethicalized awareness," by which he means a mode of life that involves appreciating nature in ways that are becoming lost, instrumentalized, and made invisible by the grid. Ethicalized awareness is, for Warner, intrinsically valuable (or something like it), but it is also of instrumental value. Ethicalized awareness is associated with the sense of agency that is important in making change. In the 1970s ethicalized awareness was central in generating consumer boycotts of CFC-propelled spray cans in the United States and Europe that were implicated in the destruction of the ozone layer.[5] And today there will not be a price on carbon in the United States unless this policy is made safe for politicians by voters signaling their willingness to accept it through their expressions of ethicalized awareness. However, ethicalized awareness is expensive in more ways than one. It is a form of attention, and attention is our scarcest resource. And sometimes change can be made and problems managed with ethical awareness playing a relatively small role. Even in the CFC case the heavy lifting was done largely though technocratic processes involving international negotiation, technology development, and the realignment of economic incentives.[6]

5. CFCs (chlorofluorocarbons) and their near neighbors HFCs (hydrochlorofluorocarbons) are fully or partly halogenated hydrocarbons that contain carbon (C), hydrogen (H), chlorine (Cl), and fluorine (F). They were widely used as refrigerants, propellants, and solvents, often marketed under the DuPont brand name Freon.

6. For different perspectives on the story of CFC management see Richard Elliot Benedick, *Ozone Diplomacy: New Directions in Safeguarding the Planet,* enlarged ed. (Cambridge, MA: Harvard University Press, 1998); Karen T. Litfin, *Ozone Discourses: Science and Politics*

This image of "management without ethics" was what many of the atmospheric scientists had in mind when they pioneered the climate change issue. The climate change story was closer to being like this (an issue the experts would manage for us) than it may seem today. By the late 1950s it was clear that CO_2 emissions put climate stability at risk. President Lyndon Johnson called out the problem in a message to Congress in 1965. Policy memos were circulated at the highest levels of the Nixon administration. Climate concerns were present in the oval office when Carter was president. In 1988 Senator Tim Wirth (D-Colorado) introduced a bipartisan bill with fifteen cosponsors calling for a 20 percent reduction in greenhouse gas emissions from 1990 levels by 2000.[7]

Suppose that as a result of these alerts and activities, a gradual transition to renewable energy had occurred, the warming had been held down, and the losers from the transition and the relatively small climate change that would have happened had been compensated. On this scenario climate change would now be fading into the background toward invisibility, a moderately unpleasant problem that would occasionally poke its head into consciousness, but one that was more or less adequately being addressed, like ozone depletion. The arguments around climate change would have been nerdy, not the ones that we have today that oscillate between denial and moral and epistemological panic. This would be a better outcome than

in Global Environmental Cooperation (New York: Columbia University Press, 1995); and Edward A. Parson, *Protecting the Ozone Layer: Science and Strategy* (New York: Oxford University Press, 2003).

7. For different perspectives on this history see Dale Jamieson, *Reason in a Dark Time: Why the Struggle to Stop Climate Change Failed and What It Means for Our Future* (New York: Oxford University Press, 2014); Nathaniel Rich, *Losing Earth: The Decade We Could Have Stopped Climate Change* (New York: Farrar, Straus, and Giroux, 2018); James Gustave Speth, *They Knew: The US Federal Government's Fifty-Year Role in Causing the Climate Crisis* (Cambridge, MA: MIT Press, 2021).

where we find ourselves in today, even though "ethicalized awareness" would not be central to the story.

But that is not the world in which we find ourselves. In any case this approach would not touch Warner's central concern about ethicalized awareness. And here Warner stands with some important figures in the American tradition of environmental thought.

American environmentalism incorporates both what we might crudely call "premodern" and "postmodern" tendencies. The premodern tendencies celebrate agrarianism, a sense of place, traditional ways of life, preservation of the world as we found it, and precautionary attitudes toward the future and technology. It is aptly expressed in the slightly dated 1979 song, "Power," that became the anthem of the antinuke movement:

Give me the steady flow of a waterfall
Give me the spirit of living things as they return to clay
Just give me the restless power of the wind
Give me the comforting glow of a wood fire
But won't you take all your atomic poison power away.[8]

By contrast, the postmodern tendency sees the environmental crisis as posing a suite of problems that can be modeled in economic terms and solved by human ingenuity and intelligence, usually as expressed in new technologies. This outlook has found expression in such movements as industrial ecology and ecological modernization and saw Buckminster Fuller and James Lovelock as inspirational figures. Perhaps it is revealing that while the premodern tendency has given

8. https://www.youtube.com/watch?v=aRMwOwo4UuY. In what can only be described as a comedic twist, the cowriter of the song, John Hall, who previously cofounded the Woodstock jam band Orleans, later served two terms as a Democratic congressman representing a predominately Republican district in upstate New York, and wrote a memoir entitled *Still the One: A Rock'n'Roll Journey to Congress and Back*.

birth to a panoply of inspirational pop songs, the postmodern turn is most powerfully expressed in popular music through irony about the premodern tendency. The Talking Heads' "Nothing but Flowers" is the magnus opus:

We used to microwave
Now we just eat nuts and berries.[9]

What the premodern and postmodern tendencies have in common is their opposition to what we might call "normal modernity," as expressed in the work of planners such as Robert Moses or corporate actors and their sycophants, who are too numerous to name. When opposing such "business as usual" practices as corporate welfare for the carbon majors, giving away the public lands to extractive industries, or permitting firms to externalize the harms they produce while privately capturing the benefits, environmentalism can present as a coherent movement. However, things can fall apart quickly when a positive vision is required.

As we saw with Stewart Brand, the premodern and postmodern tendencies can be embodied in the same person, in his case perhaps at different life-stages. In the case of the economist Kenneth Boulding, these tendencies were present simultaneously. Boulding's 1966 essay, "The Economics of the Coming Spaceship Earth," is regarded as a foundational text in the development of ecological economics, a field that is typically seen as an alternative to the conventional subfield of resource or environmental economics that emerged in the 1960s as a way of incorporating environmental externalities into economic models.[10] Yet like most conventional economic theory, Boulding's

9. https://www.youtube.com/watch?v=2twY8YQYDBE.

10. Boulding's piece first appeared in Henry Jarrett, ed., *Environmental Quality in a Growing Economy: Essays from the Sixth RFF Forum* (Baltimore, MD: Resources for the Future / Johns Hopkins University, 1966), 3–14, and has been frequently reprinted.

essay is fundamentally about identifying equilibrium states. What is innovative about this essay (in addition to Boulding's characteristically eccentric, often brilliant style, asides, and digressions) is that the search for equilibria is conditioned by seeing the earth as a closed system. But even here Boulding mentions the possibility of cracking open the closed system, at least temporarily, by developing "artificial organisms which are capable of much more efficient transformation of solar energy into easily available forms than any that we now have"—a remarkably postmodern thought for a renegade economist, well known for being a Quaker peace activist and spending his retirement years living in a collective.[11] Less well known is that Boulding was a registered Republican for most of the time he was an American citizen.

These two tendencies, the premodern and postmodern, have been part of American environmentalism from the beginning. They embody profoundly different attitudes toward science, scale, and change that can usefully be sourced to Thoreau and his lesser-known contemporary George Perkins Marsh.

Thoreau is a more complex figure than he is usually remembered, as Warner makes clear. (I marked up a draft of these comments with a number 2 pencil—as I indicated at the beginning—as a kind of homage to Thoreau the pencil manufacturer.) Still, it is undeniable that there is much in Thoreau's work and legacy that expresses the premodern tendency. He was a localist, his science was natural history, and it was seasonal and annual change that he was most interested in. While his legacy was broad and deep, it was John Muir, the founder of the Sierra Club, on whom his influence was most profound.

Marsh, like Thoreau, was a New Englander by birth, but in most other ways quite different. He spent much of his life elsewhere, first

11. Tracy Mott, "Kenneth Boulding, 1910–1993," *Economic Journal* 110, no. 464 (June 2000): F430–F444.

as a member of Congress, then five years as the US ambassador to the Ottoman Empire, then twenty-one years as the first and longest-serving US ambassador to Italy. His masterwork, *Man and Nature: Or, Physical Geography as Modified by Human Action,* published in 1864, documented, on broad temporal and geographical scales, the human transformation of the planet.

> There are parts of Asia Minor, of Northern Africa, of Greece, and even of Alpine Europe, where the operation of causes set in action by man has brought the face of the earth to a desolation almost as complete as that of the moon; and though, within that brief space of time which we call "the historical period," they are known to have been covered with luxuriant woods, verdant pastures, and fertile meadows.[12]

Even earlier (1847) he had noted the human impact on climate:

> Man cannot at his pleasure command the rain and the sunshine, the wind and frost and snow, yet it is certain that climate itself has in many instances been gradually changed and ameliorated or deteriorated by human action.[13]

Marsh thought restoration was possible through scientific management. His influence was felt most profoundly through Gifford Pinchot, the founder of the US Forest Service, Muir's great antagonist in the Hetch-Hetchy Dam controversy. Muir first saw the Hetch-Hetchy Valley in 1871 and fought to have it included in Yosemite

12. George P. Marsh, *Man and Nature; Or, Physical Geography as Modified by Human Action* (New York: Charles Scribner, 1864), 43–44.
13. "Address Delivered Before the Agricultural society of Rutland County, Sept. 30, 1847," 11. Online at https://hdl.loc.gov/loc.gdc/gdclccn.12010421.

National Park. In 1873 the renowned landscape painter Albert Bierstatdt visited Hetch-Hetchy and over the next decade produced several monumental paintings of the valley. But the city of San Francisco had the valley in its sights as the source for a public water supply that would enable growth and development and evade the private supplier, who had a monopoly. After decades of indecision and shifting decisions, congressional hearings in 1913 at which Pinchot testified in favor of the dam were decisive. The dam was approved and finally completed in 1923.[14]

While the full story is too long to tell here, the claim I want to make is that the premodern and postmodern tendencies in environmentalism reflect a tension between the following and more: appreciating nature versus using it efficiently; focusing on the local versus the global; seeing natural history as the cardinal science versus more abstract sciences that lead ultimately to earth systems science. The currently dominant outlook, which Warner identifies as "twenty-first-century environmentalism," focuses on hardware rather than software, the system rather than the behavior, and is in alliance with the grid. Indeed, global climate models divide the earth into grid cells of different sizes, often about fifty kilometers, which each cell being treated as uniform with respect to its physical properties.

Once the grid has been made visible, one sees it everywhere, and we have Warner to thank for illuminating its role in powering our devices and disempowering our ethical lives. It is this insight that leads Warner to the supposed paradox that he finds in the confrontation between environmental ethics and climate change:

14. In 1987 Ronald Reagan's resolutely anti-environmentalist secretary of the interior, Donald Hodel, shocked the conservation community by proposing that the dam be removed and the valley restored. A great deal of history has been written about the conflict over Hetch-Hetchy, but Robert Righter's *The Battle over Hetch Hetchy: America's Most Controversial Dam and the Birth of Modern Environmentalism* (New York: Oxford, 2005) is usually seen as definitive.

> In the absence of a politics for alternative infrastructure, if we look down the road that is being paved for us with green intentions, we see that we are preparing ourselves for an ever more enveloping grid modernity in which ethicalized awareness will have ever less scope. (p. 76)

There is always a hint of paradox around ethics because ethics becomes visible primarily in moments of conflict. Sometimes the conflict is intrapersonal, while other times it is interpersonal. For those of us who are pluralists about value, who believe, for example, that human well-being and natural and aesthetic values are all ultimate goods that cannot always be traded off against each other, the conflicts are more ubiquitous and troubling than for those who think that there is a single ultimate value. For such value monists there is always, in principle, some version of cost-benefit analysis that could resolve conflicts without residue. For those who think that animals and nature have serious moral value, the conflicts are even more troubling, since trade-offs between humans and the more-than human world are agonizingly difficult, often irresolvable, and almost always tragic. Into this already fraught mix comes climate change, which is the largest and most complex collective action problem that humanity has ever faced. This arrives at a time when, for deep cultural and intellectual reasons, our sense of agency is already eroding. For some people, climate change seems to finish it off altogether. We cannot help but contribute to climate change, yet as individuals we do not cause it, and so it seems we cannot stop it.

What Bernard Williams calls "the morality system" seems powerless in the face of such challenges.[15] It is understandable that we should shrivel in depression when faced with the question of how to bring about the best possible world in the face of climate change, or

15. *Ethics and the Limits of Philosophy* (London: Fontana, 1985), Chapter 10.

be reduced to tears when trying to determine our duties in the face of a global problem to which our contributions seem inconsequential.

We need to look at things in a different way. As I have argued elsewhere, a picture that views individual people as the primary protagonists—in their various roles and relationships, with their own sense of the good and sources of meaning—is one that is more natural to the climate change story than one that sees it as a problem in search of a solution from economics, technology, the behavior of nation-states, or the morality system.[16] We should return to Socrates's question, "How should one live?" If the fundamental challenge of the Anthropocene is the loss of a sense of agency, then the right answer to Socrates's question must involve the recovery (or reconception) of agency. Developing, inculcating, and acting on such green virtues as humility, mindfulness, cooperativeness, and respect for nature are important steps along the way.[17]

This thought is sometimes dismissed as proposing an individualistic solution to a collective problem, but this objection is wrong for two reasons. First, the bifurcation of actions into individual on the one hand and collective, social, or political on the other is too simplistic. Choosing the vegan option is an individual action that only infinitesimally reduces your contribution to climate change, but it is also a social action that models a behavior that is constitutive of a way of life that can make a large difference. Voting, demonstrating, or lobbying a candidate, on the other hand, are seen as political acts, but they are also individual behaviors. In any case we can work the phone bank in a political campaign and order in vegan food.

16. See, e.g., Dale Jamieson, "The Misunderstood Risks of Climate Change," *Iride: Journal of Philosophy and Public Debate* 20, no. 2 (2020): 229–36; and Dale Jamieson and Bonnie Nadzam, *Love in the Anthropocene* (New York: OR Books, 2015).

17. Jamieson, *Reason in a Dark Time*, 199–200.

The green virtues matter, not because they "solve" the "problem" of climate change through the aggregation of small actions, but rather because they help to align our actions with our values. This alignment is part of the recovery of agency, and this is what is sorely needed in the world that Warner describes. Rather than paradox, what we face is uncertainty, conflict, and even tragedy. Painful as it is, this is part of what it is to have an ethical consciousness in dark times. It may seem that we have come a long way from Warner's focus on the grid, but when the invisible is made visible, an entire world comes into view.

Can There Be a Politics of Infrastructure?

JEDEDIAH BRITTON-PURDY

What does the relationship between ethics and politics become on the grid, and what kind of politics is possible in that relationship?

The power grid is an instance of something more general. Humanity has become an infrastructure species. I mean this in several respects. The simplest is the sheer weight of our built world, its brute domination of things.

A 2017 study estimated the mass of the global "technosphere," the material habitat that humans have created for ourselves in the form of roads, cities, rural housing, the active soil in cropland, and so forth, at thirty trillion tons, five orders of magnitude greater than the weight of the human beings that it sustains. That is approximately four thousand tons of transformed earth per human being, or twenty-seven tons of technosphere for each pound of a 150-pound person. Or consider this: Human biomass in 2000 was eleven times that of all wild terrestrial mammals combined; the total weight of our domesticated animal species was twenty-four times that of the wild ones. Most of what lives here is us and our beasts.

On the Grid. Michael Warner, Oxford University Press. © Regents of the University of California 2025.
DOI: 10.1093/oso/9780197696248.003.0006

And we live in and by this heavy world. A recent serious effort to ask how many of us would survive with this sort of infrastructure stripped away came up with a global population of ten million. Increase that to a hundred million and you still have seventy-four out of every seventy-five of us disappearing.

I don't mean to put too much weight on this speculative actuarial apocalyptics. My interest is not in playing "the world without us" (or "us without the world"), but in using these infrastructural photonegatives to bring attention to how we fit into the world we have made. Our existence, as presently formed, is inseparable from the second nature that we have made out of the world, built on top of it.

This second nature is also an order, even a regime. It has a weird kind of sovereignty. It lays down the law. It says, with all the weight of the world behind it, that if you want to stay cool in the summer and warm in the winter, communicate with others, work, feel yourself a part of the cultures in which you share, here is what you must do: enter onto these roads and rails and flight routes, tap into these power grids and data networks, use these tools infused with rare earths.

As these examples show, our infrastructure doesn't dictate our ends, but it does specify our means. That is not reassuring. As I understand the ethical traditions Michael Warner is writing out of, or near, they come together in the warning that means tend to become ends, and that systems of means-deployment tend to make means even out of those who imagine they are pursuing their ends. I will come back to this point.

I have been allowing the world *infrastructure* to stand undefined, with uses clustering around Warner's grid. Let me give it a more precise sense now, and also one that goes beyond obvious family resemblance to the grid—though I will argue that it displays the same qualities that have drawn Warner's attention. I mean it to include the classic sense of roads, rails, and utility lines—artificial systems that are open to most or all, enable people to reach one another for communication or other cooperation, and can serve many possible purposes. Second, I mean those immaterial systems of interconnection and cooperation that do the same kind of work, such as the law of contract and, more generally, the legally constituted market. Although this second type of infrastructure may be immaterial, it hardly needs saying that its effects are not. Third is what we would once have called the basic domains and cycles of nature: the global atmosphere, the water cycle and the waterways that it passes through, the soil and its fertility. These, too, are the conditions of all the action and interaction we know—and, increasingly, they are things human activity has shaped and will continue to shape.

So, there is scarcely any such thing as a human being apart from a shared and artificial world: of concrete and cable and energy flows; of legal regimes for coordinating our activity, from property and contract to the Common Fisheries Policy; and of orders of power and authority, ways of generating and changing these regimes. It is within this complex housing that we variously answer the question, "What shall we do today?"

Our activity, in turn, is world-making. Answered many times over, the question, "What shall we do today?" generates the answers to other questions, "Who are we?" and "What sort of world is this?" The human species is remaking the planet as an integrated piece of global infrastructure—in its artificial carbon and nitrogen cycles, climatic patterns, energy flows, biodiversity, and skin of roads, farmland, and settlements. This fact—the collective world-making that we engage in just by living—is the starting point for the idea of the Anthropocene, the proposal that humanity has entered a new geologic epoch in which we are a force, maybe *the* force, in the development of the earth.

I have been saying "we." But as is well recognized, the Anthropocene has emerged not from a unified humanity doing something to the planet, but from some people doing things to other people, in ways that are inseparable from the forms of social order and power that have shaped the last five hundred years: imperialism, capitalism, and authoritarian socialist developmentalism among them. The *Anthropos*, the human in Anthropocene, should perhaps be renamed—hence people advocating the term *Capitalocene*, among others. All of this is true and important. I am not going to assume a *we*, an ethical subject called humanity. But I do want to inquire into the prospect of *generating* any kind of collective agency, any kind of *we*, in the face of an infrastructure world that relentlessly ties people together, however unequal and differentiating its bonds are.

* * *

As I said, infrastructure instructs its inhabitants on how to live in it, which is to say, how to live, full stop. The built world stands in the place that Rousseau assigned to his imagined Lawgiver—it changes or, better, constitutes the nature of human beings. So, an irony: On one way of telling it, political modernity comes with the discovery that our common lives have no natural form, that making them is a constructive project. The Anthropocene highlights that this is true of our relation to nature itself, that our place in the world is a political question. But the grid, the great achievement of technological modernity, seems to make a politics of these questions superfluous and elusive, even as it provides its own answers. In it, you might say, we are forced to be free—empowered in particular dimensions, with our objective, material relation to the natural world determined by the way power—in both the technological sense and in larger senses—is generated, distributed, mediated.

We will always be ruled. That is the tragic corollary of the discovery that politics has no natural form. Sovereignty may be artificial—the point of Hobbes's famous image of the Leviathan as one man made from many men—but in any tolerable human situation there will be something, someone, in the place it occupies. That is what it means to be a political species. There is a parallel point to be made about us as an infrastructure species. We will always be on one kind of life support or another; we will be one kind of cyborg or another.

What kind of political agency is suggested by our place in an infrastructure world, a grid of grids? I'd like to suggest approaching this question through the side door, by considering an arguably nonpolitical kind of agency that can feel uniquely realistic in relation to our infrastructure Leviathan. This is the hack. I mean a line of code that changes a program, or a carbon tax that changes the opportunity-cost structure of a market, or a biological tweak that makes algae a cheap way of absorbing carbon and producing fuel. It is a way of remaking a world that depends on that world's being open to revision, but, as it were, by subterfuge, almost by a kind of alchemy or magic, while the people who simply take the world for granted are sleeping, so that, when they wake up and everything is a little different, they won't even notice. It's a reformism—even a mode of revolution—entirely compatible with the kind of inattention that Warner has explored in his lectures.

It seems to me to have affinities with iconically premodern and nondemocratic forms of politics, specifically religious and courtly appeals. The hack is for our technosphere what the prayer of an adept was for seeking divine intervention: a way of getting inside the mind of the ultimate sovereign. It is a way of pursuing system-level agency in the absence of the political capacity to act at the scale that generates, sustains, and amends the system. It is a new expression of one of the oldest forms of action: murmuring just the formula that will move the mind and hand of a sovereign that is not accountable but is susceptible to an aptly phrased appeal. Once that sovereign was a king or emperor. Today it is the technosphere, the artificial system that sets the terms of all we do. It is a classic fantasy of a certain kind of intellectual that you might find just the words in your dark monk's cell and whisper them into the ear of God, or of the monarch, or of the justice holding the pivotal vote on the Supreme Court, or into the crowded void of Twitter (now X), and strike the note that resonates across the cosmos. However effective—or otherwise—this formula has been in practice before now, the tech-hack literalizes it: The chemical or software or other engineering formula really can flow into the infrastructure Leviathan's circuitry and make it cleaner, faster, cheaper—change, in other words, our infrastructure Leviathan's dictates to us, and make our infrastructure species over into a different species, one less ontologically devoted to ravaging the planet in order to live. Such a hack really may be the only readily imaginable form of agency that promises today to change the nature of human beings. This is because, as an infrastructure species, it is our nature to body forward the logic of this artificial second nature, which humans made and which makes humans.

What I have tried to do here is to follow the lines of Warner's thinking. It has led me to an image of agency in the infrastructure world that is in key ways not really what I would call political at all, that is fugitive and camouflaged, and also reserved to a certain kind of technical or administrative elite—in that sense, courtly. Electrification, Warner reminds us, consummated a certain version of twentieth-century citizenship. The electrical grid, as he evokes it, is also the paradigm of a twenty-first-century form of rule that makes a strong idea of political *self*-rule hard to achieve.

* * *

Let me name three transformations in politics, each of them exemplary of the twenty-first century even though none is exactly new. These are transformations in, or of, the classic triad of the modern state. The classic triad comprises a delimited territory that marks the scope of rule, a set of citizens who are the population of rule (both the ruled and, in democratic and republican theory and practice, the ultimate source of rule), and sovereignty, which is the distinctive form or technology of rule, a locus of the power to declare the law in that territory and for (and on behalf of) those citizens.

The first transformation is in the scope of political rule, from a delimited territory to a network of practices. We can see this pattern in transnational governance regimes such as trade law, international environmental standards, and efforts to regulate labor markets across borders—to name a few exemplars. The phenomena themselves seem to invite this redefinition of the scope of rule: They do not stay within borders, and what one has to be able to govern is the whole chain of production and transfer, the ebbs and flows of the laboring population, the health of animal populations across their migrations, or air quality as it blows across many countries. These phenomena concern linked practices across many countries.

This in turn changes the population of transnational governmentality from citizens to stakeholders. The question is not the color of your passport but your involvement in this network of practices. This status cross-cuts national citizenship, as the practices cross-cut borders.

Third, sovereignty is replaced by governance, the process generating the norm that shapes the system. Sovereignty centers on the locus of decision, and by that fact implies that the citizens of a polity may *choose* the terms of their common lives, and so the shape of their future. This is an elusive idea, even a regulative one in Kant's sense—not to be literalized in naive fashion, but to be taken as a principle that guides assessment of actual institutions. In this sense, the distinction between sovereignty and governance steps into the light. An election is a mode of sovereignty. So is a less democratic form of choice, down to the extreme of an autocratic decree. By contrast, a market cannot be a mode of sovereignty, but its structured dispersal and ordering of decisions is a mode of governance, a source and style of order. Governance may take many forms: administration by experts who deny that they are engaged in substantial choice rather than being guided by truth; negotiation by representatives of interest groups; even the mute guidance of technology and infrastructure.

For the moment, I want only to observe the tendency to talk of grids in the same way I have been sketching: as administered networks with stakeholders whose interests have to be taken account of in some enlightened way but whom it would be a sort of category mistake to imagine as involved in self-rule with respect to the grid. We seem to have come to a paradigm of contemporary governmentality in the grid.

Here is a second affinity: between the economy of attention that the grid cultivates, the obliviousness of participating in it, and the economy of attention in the price system that is a main mode of governance as well as the central mechanism of economic coordination in market society. The great neoliberal defense of markets, in Friedrich Hayek's writing, does not depend on fairness, desert, or even any very robust idea of efficiency, but rather on the idea that pivoting all our actions around prices lets us act as if we knew a great many things without knowing them at all. If you go to Home Depot to pick out kitchen tiling and the copper tiling has gone up by a dollar, you will choose as if you knew the reasons—as if you knew there is political instability in some copper-mining regions or a new technology that will demand copper for some lucrative use—and so your decision will approximate a certain kind of globally rational allocation, even as you are only aware of looking at a price tag that says $1.79 and another that says $3.79.

We are in Thoreau's dystopia here. Maine and Texas are connected, and not only do they have nothing to say to each other; they do not even need to know they are connected. (Strictly speaking, Texas's grid is an isolate, but that fact does not defeat the main point.) Market governance has its own ends—that resources should go to those who command the greatest purchasing power, rather than serve any other notion of human need or deservingness, and that the disposition of resources should maximize financial return on investment. Yet no one within this economy of attention is obliged to pay attention to these purposes, to take them as actionable questions rather than unquestionable facts.

Governance eclipsing sovereignty, stakeholders replacing citizens, networks overgrowing territories: these forms of ordering, with their affinities to the grid, are forms of antipolitics. Antipolitics is a paradoxical form of politics that takes questions off the table of collective contest and decision and moves them into more opaque processes, making them someone else's question—an expert's, an industry's, and so on—rather than a question for each citizen. I think it is a part of reflectiveness in modern life to be ambivalent about these forms of depoliticization, to be liberated by them from certain burdens and fears, and also to be haunted by awareness of falling into irresponsibility, of becoming the tools of our tools, and becoming other people's tools (or the tools of their tools), precisely as Thoreau warned.

What are our alternatives? I find it sobering that the grid, huge and important as it is, is also in a way only a synecdoche for these other forms of governance in which we find ourselves. What we are exploring is very large indeed. I find it sobering, too, that the most ambitious contemporary effort to set out an ethics of attention, knowledge, and responsibility, in the Kentucky writer Wendell Berry, concludes in a very strong localism. If we are going to know what we depend on, what harm we do, what we can take care of, Berry argues, the scope of our actions and our economic entanglements cannot extend far beyond the horizon of our sight, the ridges that bound our watersheds, our concrete personal relationships. I confess that I think if this conclusion is right, it is a counsel of despair: Localism is far from being a viable option for people whose entanglement in global networks of market governance has greatly intensified in the last fifty years. It is, alas, a luxury.

And the localism that Berry advanced is not just about scale but about the marriage of scale and responsibility—not having your own little home coal mine, as some farms did in Western Pennsylvania where my family used to be, but something closer to the methane digesters that my neighbors went to China to learn to build when I was growing up in West Virginia, among people who were influenced by Wendell Berry. The call is really revolutionary: to withdraw from entanglements with our vast systems of energy, work, production, and consumption while replacing them, alone or with neighbors, with alternatives that are qualitatively as well as quantitatively different. And of course such heroic efforts cannot really be alternatives to large-scale politics because the viability of the alternative depends in key ways on what is happening in the main system: solar production on Chinese and now US industrial policy, local agriculture on the pace of climate change, and so forth. Attempt to leave behind the large and commensurately opaque systems of twenty-first-century sovereignty and governance and one is drawn ineluctably back into them.

This paradox calls me back to Thoreau's view of citizenship, which is defined by complexity and ambivalence. Bound up with other people, he became implicated in their complacency, their imperialist and racist arrogance—and, among the abolitionists of his family and friends, in a kind of meta-problem, their self-righteousness and self-congratulation about being better than their complacent neighbors. I suspect the shape of the experience does not seem entirely anachronistic, if you have been awake in recent years.

He saw that there is no getting away from being trapped in a community of purposes and ends—no matter how bad and unwelcome they are. But as long as there is politics, being trapped need not be the end of the story. A political community is also one in which it remains possible, as Thoreau put it in "Civil Disobedience," to appeal against the people to themselves—to make the claim that we could do, and be, something else, that we carry in ourselves the possibility of a different world.

The enemies of that possibility include inattention and its systemic corollary, the immurement of purposes and biases so deep in the fibers of a governance regime that one can live by them without ever recognizing them. The enemies of possibility also include the individualistic and voluntaristic conceit that a few extraordinary individuals, leading exemplary lives, can lead the rest of us to a better way. This may be true in matters of inner conversion, but it cannot be enough for questions that center on the terms of our material interdependence, the economic and energy systems that hold us together and make and remake our world. What we still need for a politics of our grids, and of our infrastructure at large, is a way of combining the vividness of vision that dissenting prophets and communities offer with the reach and power that only sovereign action can achieve. Michael Warner's lectures call us back from our inattention to fresh engagement with this most difficult of problems.

Grid, Power, Poetry

ANAHID NERSESSIAN

PROLOGUE: AT THE END OF FIRST LIGHT

As it was printed in 1939, Aimé Césaire's *Notebook of a Return to the Native Land* begins with an incantatory farewell to the very long age of Enlightenment, that stretch of time that, in his telling, began its ingenious, bloodthirsty run in or around 1492. Historians usually confine the Enlightenment to the roughly one hundred years between the late seventeenth and late eighteenth centuries, assigning similarly descriptive names—the Age of Exploration, the Scientific Revolution, the Industrial Revolution, the Age of Empire—to the ones that bookend and overlap them. For Césaire, however, each of these eons is subordinate to one, whose self-identification with the figure of *light*—standing aspirationally for the values of discernment, insight, and what Immanuel Kant calls "man's deliverance from his self-imposed puerility"—the *Notebook* flips on its head.[1]

1. Immanuel Kant, "Beantwortung der Frage: Was Ist Aufklärung?," in *Kants gesammelte Schriften*, vol. 8 (Berlin: Georg Reimer, 1910), p. 36. My translation. As readers of Césaire will know, subsequent editions of the *Notebook* begin with the phrase "Au bout du petit

On the Grid. Michael Warner, Oxford University Press. © Regents of the University of California 2025.
DOI: 10.1093/oso/9780197696248.003.0007

The Age of Enlightenment was, after all, the Age of Enslavement, and it laid the ground for the immiserated conditions of colonial society the poem so vividly indicts. When the *Notebook* begins, the Enlightenment is on its way out. Its demise doesn't solve any problems; too much of a mess is left behind. Still it is important to Césaire to tell us exactly where we are before we can even begin to figure out where we are going. "Au bout du petit matin," he writes, not once but near two dozen times. "At the end of first light" many things become visible, first and foremost "the hungry Antilles, the Antilles pitted with smallpox, the Antilles dynamited by alcohol, stranded in the mud of this bay."[2]

In the next several strophes, the view progressively widens until we have moved from "the dust of this town sinisterly stranded" to an ominous wide shot of the whole world and back again:

> At the end of first light, on this very fragile earth thickness exceeded in a humiliating way by its grandiose future—the volcanoes will explode, the naked water will bear away the ripe sun stains and nothing will be left but a tepid bubbling pecked at by sea birds—the beach of dreams and the insane awakening.
>
> At the end of first light, this town sprawled—flat, toppled from its common sense, inert, winded under its geometric weight of an eternally renewed cross, indocile to its fate, mute,

matin" and are followed by an apostrophe to the colonial figure of the "flic" and evil "gris-gris"—"Va-t-en, lui disais-je, gueule de flic, gueule de vache, va-t-en je déteste les larbins de l'ordre et les hannetons de l'espérance"—before returning to repeat, as in the original text, the spatial and temporal mantra of "au bout du petit matin." See Aimé Césaire, *Cahier d'un retour au pays natal* (Paris: Présence Africaine, 1983), 7.

2. Césaire, *The Original 1939 Notebook of a Return to the Native Land*, bilingual ed., ed. and trans. A. James Arnold and Clayton Eshleman (Middletown, CT: Wesleyan University Press, 2013), 2–3.

> vexed no matter what, incapable of growing according to the juice of this earth, encumbered, clipped, reduced, in breach of its fauna and flora.
>
> At the end of first light, this town sprawled—flat.[3]

What does this apocalyptic sequence suggest? That empire's brave new world is gone but its harms remain. The end of first light cannot therefore be the dawn of a new beginning; it rather announces the closure of a set of options for human life and civilization. Now in the midst of an "insane awakening," we find ourselves in the midst, too, of cascading social and ecological disasters we have no idea how to fix, poisoned by a history so noxious it seems already to have contaminated any possible future. And yet it is here that we must begin.

Michael Warner's lectures begin, too, at the end of first light, both because they must and because the intellectual trajectory they map—from Foucault's theory of power to the ethical crisis triggered by the powering of the electrical grid—includes a stark confrontation with the legacy of the Enlightenment, from which it is impossible to separate our persistent demand for more, faster, better, and climate-friendly light. Warner phrases this confrontation in terms of an apparent impasse between the need to decarbonize the grid by transitioning to renewable energy and the necessity of understanding ourselves as moral agents, who ought to want and demand more from such a transition than that our lives be minimally interrupted.

The question of the grid, as Warner poses it, is really this: What world do we want at the end of first light, and what kind of species do we want to be? Do we want to hang on to a vanishing world of ease and plenitude for some, at the price of unconscionable suffering for most, or do we want radically to reimagine what is meant by human

3. Césaire, *Original 1939 Notebook*, 3.

flourishing beyond the limited, exploitative, and wholly unsustainable contours of the grid as it presently exists?

The grid, Warner explains, begins with the city. It "has a long history of association with state power, stretching from Babylon to the Roman, Chinese, and early modern Spanish empires" (p. 36), but in North America it begins with colonialism and the founding of East Coast urban towns like New Haven, Philadelphia, and Savannah. In "London," a poem about another city, William Blake called the grid-based process of "commensurating space into property" chartering, as in

> I wander thro' each charter'd street,
> Near where the charter'd Thames does flow.
> And mark in every face I meet
> Marks of weakness, marks of woe.
>
> In every cry of every Man,
> In every Infants cry of fear,
> In every voice: in every ban,
> The mind-forg'd manacles I hear.[4]

Thus the process of standardization—a favorite desideratum of Enlightenment thought put into practice—is also always a process of privatization, introducing a scheme of arbitrary and phantasmatic value where there was once merely a resource. "To charter" means all at once "to map," "to hire," and "to empower." It is the action of the state and the corporation, but it can also be—for better and for worse—the action of the poetic line, itself as much a "mind-forg'd manacle" as the grid that controls how you walk the street and where

4. William Blake, "London," in *The Complete Poetry and Prose of William Blake*, ed. David Erdman (New York: Anchor Books, 1982), 26–27.

the river runs. (Later in life Blake would state unequivocally that "Poetry fettered, fetters the human race.")[5]

In this response to Warner I'll present a counterpoetics of the grid, defined as a cultural and political consciousness that arises not exactly in opposition to gridded modernity but that nonetheless challenges its conditions by *unchartering* the grid's rhetorical and representational forms. This counterpoetics can be found in poetry—and I find it there, below—but it can also be found more widely in modes of thought that seek out and articulate alternative styles of living with others. Such styles may serve, in Foucault's words, "as functional devices that [might] enable individuals to question their own conduct, to watch over and give shape to it, and to shape themselves as ethical subjects; in short, their function [is] etho-poetic."[6] For all this talk of "individuals," these devices' ultimate purpose is to reorient the self toward the collective. If they cannot by themselves dismantle "the framework of individual moral choice," they might nonetheless socialize it. Or try.

The rest of this essay will proceed as follows. In the first main section, I'll follow Warner in considering the grid's implications for moral philosophy from a variety of perspectives. In the second, I'll turn to poetry as well as some works of visual art in search of a resistance to the ethical foreclosure that, paradoxically, threatens to accompany the most pressing ethical—not to say existential—problem of the twenty-first century, namely how to adapt to the exigencies of climate change without intensifying social inequity. A more utopian way to phrase this would be to ask how such an adaptation might be just, how it might force us to abandon the social and economic systems that have brought us to the brink and establish new ones in their

5. Blake, *Jerusalem the Emanation of the Giant Albion*, in *Complete Poetry and Prose*, 146.
6. Michel Foucault, *The Use of Pleasure*, vol. 2 of *The History of Sexuality*, trans. Robert Hurley (New York: Vintage, 1990), 13.

place. While there is not enough space in the following discussion to address this particular matter head on, it should be understood as the horizon of all ecological inquiry worthy of the name.

ETHICS OFF, ON, AND BEYOND THE GRID

The problem with simply going *off* the gird, Warner notes, is that it tends to entail a rejection of collectivity. Communes can be wonderful, but the future of the planet rests on transformations of an extraordinarily vast scale, not on midsized groups of like-minded people going into the woods because, like Thoreau, they wish to live deliberately. On a more sinister note, the offgrid mentality is often very close to the survivalist or "prepper" one, with all the unsavory, often white-supremacist politics that tend to accompany that disposition. Regardless, whatever its ideological basis, going offgrid stages a melodrama of refusal that is often profoundly antisocial.

Indeed, this was the complaint registered by the journalist Kathryn Schulz in her 2015 hit piece on Thoreau in the *New Yorker*, in which *Walden* appears "less a cornerstone work of environmental literature than the original cabin porn: a fantasy about rustic life divorced from the reality of living in the woods, and, especially, a fantasy about escaping the entanglements and responsibilities of living among other people."[7] To be clear, I find this account of *Walden* hollow and misleading, but it's a useful measure of the gap between how we think about effective environmental action and what effective environmental action actually entails. Consider the word *porn*. For Schulz, Thoreau is a bad environmentalist because—whatever his actual practices—his imagination is insufficiently ascetic, even

7. Kathryn Schulz, "The Moral Judgments of Henry David Thoreau," *New Yorker*, October 12, 2015. Online at https://www.newyorker.com/magazine/2015/10/19/pond-scum.

obscene. He enjoys his isolation and abstemiousness a little too much; he claims to abstain from plenitude but recovers it in his fantasy life. The implication here is that scarcity, of both the material and the mental kind, is the proper arena of morality. It's not enough that Thoreau live alone in his cabin with his table, desk, bed, and three chairs. He can't enjoy it; that would be cheating.

The problem with this way of thinking is simple. From an ecological perspective, it does not matter at all what any *one* of us does, including Thoreau. It does not matter that Thoreau's pleasure in his renunciation of middle-class New England life is so intense it seems indecent. It does not matter if I recycle today or not, and it matters even less if I enjoy it. More to the point, the grid will be decarbonized not because individuals do or do not accept "the entanglements and responsibilities of living with other people." In all likelihood, it will be decarbonized because the relevant corporations will have figured out how to make the transition profitable.

As for entanglements and responsibilities, the grid effectively erases all trace of them, as it erases all trace of the enormous labor force engaged in fueling and maintaining the grid itself: miners, engineers, computer technicians, and so on. Allison Carruth, in an essay on cloud computing that speaks just as well to obscurities of the grid, writes that behind our "cultural values of connectivity and speed," behind "every stroke of the keyboard and swipe of the touch screen," there is "the meat"—the bodies, machines, and materials that make everything happen, and that we notice only when the frictionless experience of sending this or buying that is unexpectedly disrupted by a server crash or strike.[8] The problem posed to the sociality of the future by the electrification of everything is precisely that the grid, in its ideal form, makes all our interdependencies—all our

8. Allison Carruth, "The Digital Cloud and the Micropolitics of Energy," *Public Culture* 26, no. 2 (2014): 343–44, 360.

meat—disappear. You might even say that, in the grid as in all other commodities, "a definite social relation between men" assumes "the fantastic form of a relation between things."[9]

The protagonist of this story is the consumer whose maximal demand for power is matched by a minimal awareness of all that such a demand asks us to give up—including, as Warner says, our potential to be and act as ethical agents. Modern mainstream environmentalism catechizes us in routines of restraint and self-abnegation—reduce, reuse, and recycle, ad nauseum—but the environmentalism of the future may commit us by contrast to a seamless integration of the needs of the person with the load of the system.

To what end? One dystopian possibility would be the subsumption of the human category of action by the techno-materialist category of usage, with all sorts of dubious consequences for our notions of morality, accountability, and social obligation. Moreover, if Anthropocene Now involves the same metamorphosis of natural resources into market commodities proper to the carbon-based economy of the Industrial Revolution, Anthropocene 2.0 hopes not only to abandon coal but to transition away from a division of "nature" and "market" altogether. What is already there—wind, solar, tidal—will be there for us to spend. What we will spend will be there already for us.

In theory, the naturalization of the grid makes it amenable to any number of political platforms. As Warner has it, "Infrastructural solutions to greenhouse gas emissions . . . can be applied in authoritarian cultures as easily as in democratic ones, if not more so" (p. 50). In practice, I think this antinomy may be overstrict. It goes without saying that there exist formal democracies that make authoritarian use of their utilities, from water to power to prisons and the police. Take, for just one historical example, the forcible displacement of Indigenous

9. Karl Marx, *Capital*, vol. 1, trans. Ben Fowkes (New York: Penguin, 1990), 165.

people by the Army Corps of Engineers during the implementation of the Pick-Sloan Plan. Under the aegis of providing hydroelectricity to the prairie states, the Corps invoked eminent domain to seize 550 square miles of Native land along the Missouri River, driving 30 percent of people living on reservations away from their homes.

Nick Estes, from whom I take this example, calls the Pick-Sloan Plan "a twentieth-century Indigenous apocalypse" and links it directly to the construction of the Dakota Access Pipeline—also the work of the Corps of Engineers—and the violent suppression of the #NoDAPL protests by the US government.[10] Elsewhere, Estes has warned that the push for a clean-energy transition may once again expedite the ongoing appropriation of Native land and the ongoing dispossession of Native people. He has observed, too, that the decentralized nature of solar and wind power makes it far more amenable to communal stewardship and regulation than, say, nuclear power, which has so many risks that a hierarchical management structure is necessary just to keep plants—and people—safe from catastrophe. By contrast, says Estes,

> You knock a couple wind turbines off the grid and it doesn't have any effect. . . . And even the fossil fuel industry is thinking about this. They want to recentralize those decentralized green technologies. It's like the internet: everybody thought it was going to democratize everything, and now it's been totally privatized and commodified. That's something we have to fight in this energy transition.[11]

10. Nick Estes, *Our History Is the Future: Standing Rock Versus the Dakota Access Pipeline* (London: Verso, 2019), 134.

11. Estes, "Water Is Life: Nick Estes on Indigenous Technologies," *Logic Magazine*, Issue 9 (Dec. 7, 2019). Available at https://logicmag.io/nature/water-is-life-nick-estes-on-indigenous-technologies/.

In other words, the structural resilience of renewable energy—combined with a political awareness of what kinds of human and natural disasters infrastructural work has precipitated—might help us evolve an alternative to the authoritarian/democratic binary. They might, that is, make possible a new social, economic, and ecological model in which our relationships to one another and to the land are not occluded by technology but used to negotiate it, and in which the collective might supplant the individual as both subject and object of ethical action.

To summarize: For Warner, the crisis posed by the grid is essentially ironic and potentially tragic. It doesn't just challenge our customary ideas about everyday environmentalist practice, it upends the entire premise of classical Western ethics, creating conditions whereby an adaptation necessary for our survival as a species may all but eliminate our ability to imagine ourselves as moral agents. The grid's picture of a sustainable future is one in which it does not matter how much you take because you are never directly taking from others; this sounds very well, but it also risks the disintegration of the entire self-other relation, which is to say the core of both practical and normative ethics as we have long if not always understood it. This in turn has political consequences. Will griddedness take us back to the frontier and its disastrous, indeed genocidal, vision of every man for himself, or will it take us forward into the kind of ecologically oriented collectivity Estes describes? This language of backward and forward is metaphoric: Both vistas are visible ahead. Both possibilities are on the table.

Here is something we have yet to bring into the conversation. It's that, when we're talking about ethics, we're really talking about something exceedingly simple and mind-blowingly complex. We're talking about human flourishing, and what constitutes the best, fullest, and most just expression of our capacities. Foucault thought about the concept of flourishing for the entirety of his career, often locating it behind a variety of prisms: the institution, sex, the self, the state. But it has been asked for a very long time. It comes to us at least from Aristotle, and therefore belongs to a premodern conception of the human mediated by its relation to the species and not by any notion of contract, care, or control. In other words, it belongs to a classical account of ethics that does not focus on the relation of the self to the other but the relation of the self to its own species-being, which is to say, of the self to its proper expression as a member of a group.

What threat does the grid, with its explosion of an ethical ideal grounded in interpersonality, pose to our species-being, or rather to its development? How might it actively contract and warp the scope of what we are and what we ought to be? We have already seen that, despite its projection of absolute fluidity and its concealment of human labor, the grid depends on work, and the work is done by people. Engineers design our grids; technicians build and fix them; electricians run our wires; a recent report from the Tech Transparency Project suggests that Apple, which has been building wind farms in Xinjiang, is relying on the forced labor of the Uyghurs to get their grid up to speed.[12] It may encourage fantasies of an automated and, indeed, a transubstantiated future, but the historical incarnation of the grid is, in a word, carnivorous. The line between grid and gulag is thin.

The language of transubstantiation is borrowed from Marx, who, in the first volume of *Capital*, explains that in order to "operate effectively as exchange-value," the commodity "must divest itself of its natural physical body" and undergo an "act of transubstantiation" more extreme than "the casting of his shell for a lobster."[13] The grid must do

12. "Apple's Uyghur Dilemma Grows," *Tech Transparency Project*, June 8, 2021. Available at https://www.techtransparencyproject.org/articles/apples-uyghur-dilemma-grows. Last accessed June 18, 2022.

13. Marx, *Capital*, 197.

the same. As Warner so beautifully puts it, "The electric power grid is more than an object of forgetting; it is an engine of unconsciousness" (p. 30). To put these ideas together, we might say the grid imposes a metamorphosis on our ethical self-awareness, so that we do not merely forget the existence of other people, we forget the existence of *their bodies*. We forget that it is "human brain[s], nerves, muscles, and sense organs" making and disappearing beneath wires and wind turbines, behind dams and pumps and chips and screens.[14]

What would it look like to remember an embodied relation to others? I don't think anyone really knows, though plenty of people have ideas. Marx famously called on the notion of species-being to evoke a world in which "the natural physical body"—the body of the species—might define the social relation and vice versa. In his "Comments on James Mill," a document often derided as a relic of the humanist as opposed to the scientific Marx, he predicted four consequences of such an existence. First, an ecstatic continuity between what I make and who I am. Second, the guarantee that whatever I make satisfies an essential human need in others. Third, the emergence of a new sociality in which I am, at last, the extension and completion of your essential nature, and you are the extension and completion of mine. And four, a materially metaphysical concord between the "individual expression of my life" and the individual expression of yours such that "in my individual activity I [will] directly confirm . . . and realiz[e] my true nature, my human nature" as "communal nature."[15]

This is intimate, even erotically charged language for a nineteenth-century economist. Perhaps it's even a bit Thoreauvian. It expresses the desire for a corporeal technology through which labor itself might

14. Marx, *Capital*, 164.
15. Marx, "Comments on James Mill," excerpted in *Karl Marx: A Reader*, ed. Jon Elster (Cambridge: Cambridge University Press, 1986), 34.

be literally reincarnated, a process of counter-transubstantiation that might give our bodies back to us at last. Such a process cannot occur without or in despite of other technologies—this is perhaps the greatest insight of Marx's historical materialism. Thus against any intuitive opposition of infrastructure and embodiment, and against the fantasies of forgetting and unconsciousness the grid encourages, we ought to say: The grid already belongs to us because we are inside it, like a virus within a host.

OTHER GRIDS, DIFFERENT FUTURES?

The second half of this short essay asks what we can discern of our relation to the grid from literature and art history. I have already proposed the existence of a counterpoetics of the grid based on Foucault's notion (itself imported, as Foucault says, from Plutarch) of etho-poetics, or the ethical crafting of the self through habit, ritual, and repetition. A different way of framing this counterpoetics would be to invoke another Foucauldian notion, this one drawn from his 1968 tribute to Jean Hyppolite, who had been his teacher at the École Normale Supérieure and whom Foucault replaced following Hyppolite's sudden death.

There, Foucault describes philosophy as being in an agonistic relationship with its own "nonphilosophical excess," which, in Lynne Huffer's helpful formulation, resists "a rhetoric of persuasion that manipulates and polices under the blinding glare of Enlightenment surveillance" to serve, instead, "as a small, humble 'light which kept watch'" over alternate practices of thought, action, and language.[16] Contrast this "light which kept watch" to the light that Césaire begins

16. Lynne Huffer, *Foucault's Strange Eros* (New York: Columbia University Press, 2020), 33.

to dim at the opening of his *Notebook*. That light, as we've seen, was also "the blinding glare" of the Enlightenment, and it is opposed by an anticolonial poetics that celebrates

> those who invented neither powder nor compass
> those who could harness neither steam nor electricity
> those who explored neither the seas nor the sky
> but knew in its most minute corners the land of suffering.[17]

This is, I suggest, a poetics deeply committed to keeping watch over a collectivity that is, by its very nature, opposed to the antiethics of the grid. Where else might we find it?

Consider the work of Bay Area poet Juliana Spahr, who has described her own engagement with ecological themes in part as a critique of traditional "nature poetry," with its pastoral isolation of the environment from society and technology. In its place, Spahr proposes "a poetics full of systemic analysis that questions the divisions between nature and culture," going so far as to call poetry that claims to be "green" but fails to trouble those divisions immoral. More to the point, and as she says in her 2011 collection *Well Then There Now*,

> I was . . . suspicious of nature poetry because even when it got the birds and the plants and the animals right it tended to show the beautiful bird but not so often the bulldozer off to the side that was destroying the bird's habitat. And it wasn't talking about how the bird, often a bird which had arrived recently from somewhere else, interacted with and changed the larger system of this small part of the world we live in and on.[18]

17. Césaire, *Original 1939 Notebook*, 35.
18. Juliana Spahr, *Well Then There Now* (Jaffrey, NH: Black Sparrow, 2011), 69.

Part of Spahr's response to this suspicion was, as it happens, to enroll in an ethnobotany course at the University of Hawai'i, where she was at that time employed. After all, she writes, "Poets need to know the names of things"; she wanted to learn the names of things that grew in Hawai'i.[19]

Spahr's 2015 poem "Turnt" is not about Hawai'i, but it is about names. It is set during the Occupy movement (2011–12) and describes the heady days of what Spahr elsewhere calls "almost-revolution," a volley of riots and occupations that never develops into a concrete social movement and whose political effects are uncertain. As Spahr describes the action of the crowd as it moves through city intersections—shades of Blake's "charter'd streets"—it takes the reader some time to notice that she is incorporating proper names into her lines in a loosely alphabetical order: "Someone grabbed me from behind and I thought it is Artem but later / realized it was Berat. . . . I grabbed Charlotte's hand and held it for a while when things felt dicey"; "Emma there, throwing eggs"; "I love you I texted Felix/Lub u!!!!!! I texted Haruto"; "Miguel stays home with Minjoon"; "Mohamed, my fighting teacher, fights."[20] Here is a longer passage:

> I don't want it to be heroic but last night I turned the corner and Nor was there with her bike and when I saw her I said I love you and we walked down the street as each window was cracked. They got turnt. Eventually we disperse. I jog for a few minutes away and out of the kettle. We joke, circle back to watch a car burn. Oliver walks by. He is hurrying towards the dispersal. I love you we say to him as

19. Spahr, *Well Then There Now*, 70.
20. Spahr, "Turnt," in *That Winter the Wolf Came* (Oakland: Commune Editions, 2015), 84–86.

> he heads off. The car burns. The fire truck arrives. As I stand there watching it, it is as if everyone I have ever texted I love you to walks by. I love you we call out to each other.[21]

What Spahr does not advertise until the very end of "Turnt," when she makes it explicit, is that the poem has been laboring under a quasi-Oulipian constraint. "This poem is true," she tells us, "But the names are not true":

> I found a list of the most popular baby names for various countries in 2015, the year in which I am writing this poem. I made a list, one male and one female from each list. Then I alphabetized it. And I put those names in this poem one by one. I got to O.
>
> But Olivia, Saanvi, Santiago, Seoyeon, Sofia, Yui, and Zeynep, I love you too.[22]

Imagine, for a moment, that the lines of the poem form, as they do quite literally on the page, a train of horizontal lines. Imagine those names—*Artem, Berat, Charlotte, Sofia, Yui, Zeynep*—falling into the vertical line of Spahr's alphabetized list. The lines move into the names and the names fall into place, their crossing or confluence named "love." They are on a grid.

It's no accident that Spahr, who always works cleverly and tenderly with vernacular culture, chooses "the most popular baby names for various countries" to be the bodies on the line in this movement. You might say these names are basic; you might say that they are exemplary, that they hold the most people without holding everyone,

21. Spahr, "Turnt," 87.
22. Spahr, "Turnt," 87.

because this is not a poem for everyone. It is, in many ways, a hard poem, and its ethics are hard, too. They are directly engaged in thinking about how abstract forms of organization can be made—just like the grid, just like our relation to any resource—ethical as well as emancipatory, serious as well as sensuous.

Warner's lectures are not, of course, about poetry. But they are about embodiment or, to use an old theological term, enfleshment, and about how concepts are transformed into instances: how energy becomes a resource, power becomes a utility bill, the human being becomes the consumer. Art is excellent at mediating between these two domains or categories (concept and instance) while also suspending or holding off the complete dissolution of theory into practice. That's why I'm turning toward it, to provide a perhaps unlikely perspective on the grid as a technology that likewise toggles the abstract and the material. My purpose is twofold. First and most simply, I want to put some additional terms on the table for thinking about structure, shape, and materiality; second, I want to ask what a poetics responsive to the condition of the worldwide grid might look and sound like, with one foot in the grid's history and another reaching, however tentatively, toward its future. Despite rumors to the contrary, poetry is not prophecy. Still, it might suggest refinements of the future's preliminary forms in the present, including challenges to the vanishing horizon of interpersonal ethics Warner so movingly describes.

One might think, in this imagined or propositional lineage, of the work of Bernd and Hilla Becher, the German husband-and-wife team of conceptual artists and photographers best known for their photographs of industrial buildings or objects.[23] These photographs, which the Bechers referred to as "typologies," consist of stark infrastructural

23. For some examples of their photographs, see https://www.sfmoma.org/exhibition/bernd-hilla-becher/.

portraits of water towers and gas tanks, coal bunkers and blast furnaces, grain elevators and lime kilns, and high-tension electrical pylons sorted into grid formations comprised of vertical and horizontal lines. When the Bechers put the grid to use in this way, it was in response to a much older and much contested modernist tradition, the sort we would associate with Malevich, Mondrian, Agnes Martin, Sol LeWitt, Mary Heilmann, and, more recently, with the ominous post-pop art of Sarah Morris or the deceptively playful psychedelics of Maya Hayuk.

In her essay "Grids," published in *October* in 1979, Rosalind Krauss finds the grid "announc[ing] . . . modern art's will to silence, its hostility to literature, to narrative, to discourse" and, with them, to history.[24] The artist and critic Andrew Menard puts it similarly: "The grid is an attempt to overcome history by design."[25] For Kraus, the grid too stands against history as development, in a way that makes it seem like the proleptic aesthetic analogue of what someone will eventually call the end of history. "Antinatural, antimimetic, antireal[,] [i]t is what art looks like when it turns its back on nature" and, with nature, the emancipation of life through the fulfillment of historical processes.[26]

In a moment that resonates nicely with Warner's remarks on the infrastructural unconscious, Krauss writes that if the grid is against history, it is better disposed to etiology, even to the point of substituting the former with the latter. Here is Krauss:

> History, as we normally use it, implies the connection of events through time, a sense of inevitable change as we move from one

24. Rosalind Krauss, "Grids," *October* 9 (Summer 1979): 50.
25. Andrew Menard, *Sight Unseen: How Frémont's First Expedition Changed the American Landscape* (Lincoln: University of Nebraska Press, 2012), 115.
26. Krauss, "Grids," 50.

> event to the next, and the cumulative effect of change which is itself qualitative, so that we tend to view history as developmental. Etiology is not developmental. It is rather an investigation into the conditions for one specific change . . . to take place. In that sense etiology is more like looking into the background of a chemical experiment, asking when and how a given group of elements came together to effect a new compound or to precipitate something out of a liquid. (64)

In the grid, in each meeting of abscissa and ordinate, there is what Lacan would call a *point de capiton*, a quilting-point whose job is to hold together the facticity of matter to the desire that it signify something beyond itself and something beyond us. The electrical grid does this nicely. It is invisibly everywhere, its appearances largely private and indirect—the flip of a switch, the wrenching of the toaster's dial—and it claims the power to save us from ourselves.

Warner suggests as much when he invokes the unconscious by way of situating our epistemic grasp on griddedness. For Lucy Lippard, whose well-known critique of postmodernist art for its persistent dematerialization is relevant here too, the art-historical grid represses the historicity of the industrialized landscape, borrowing its "back roads" to simulate the "invasion of previously wild . . . places" while also rendering that invasion inscrutable.[27] This grid reliably stipulates itself as flat; so does the ego, whose longing for an unbroken contour of self and intention forces depth to cave and surface.

The Bechers' work, however, does not comply with these requirements. For one thing, its documentary ethos is plainly historical: It

27. Lucy R. Lippard, *Undermining: A Wild Ride Through Land Use, Politics, and Art in the Changing West* (New York: New Press, 2014), 17.

is overtly focused on the means of industrial production and alludes to industrialism's productive modes, as well as their obsolescence in a postindustrial moment. For another, the elaboration of the line into the photograph turns the modernist grid into something almost baroque, the play of light and shadow in these black-and-white images creating the effect of information being lost even as so much is being preserved, of excess uncontainable by system and yet part of the system, too. We should keep in mind, on this front, that the Bechers started out referring to their photographs as sculptures, as in the title of their debut photobook, *Anonymous Sculptures,* published in 1970.

The avalanche of attention paid, in recent years, to the Bechers' project is perhaps a sign of the infrastructural unconscious finding its extimate expression. It also raises the question of what kinds of politics attaches itself to an aesthetics that is so explicitly engaged in making conspicuous the nearly invisible work of powering society. If people are interested in the Bechers they are interested in this question, too, and in what we might call the question's beauty, in Kant's sense of the beautiful as a propaedeutic for moral feeling. In one version of a best-case scenario, seeing the objects that powered the old grid, or in which the old grid's potency was stored, might enlarge our sense of what a differently gridded future might make possible.

This is a pretty social-democratic *might* and a pretty social-democratic *possible,* since it remains attached to both the past and the prospect of state-run infrastructural projects. The Bechers' work is often described as dystopian, but in it I see a potent nostalgia for the so-called European social model, for a strong welfare state motored by manufacturing and buoyed by a certain cultural respect for the worker and his domain. Their aesthetic is one perfectly in sync with the philosophical model proffered by Jacques Rancière, with his now well-worn paradigm of art and politics as varying distributions of the sensible; even the language of distribution implies networks of transport and usage and, as in the Bechers' photography, an apostrophe to some centralized agency, a plea that it responsibly and equitably pass out its goods and services. If Rancière has become necessary to discussions of contemporary art, that may be because his is a rhetoric of necessities mediated, in one way or another, by the state. In an age where the state seems so indifferent to the basic needs of persons and systems, it's not surprising that it would become the object of robust fantasy and longing.

Rancière asks art for a simple favor: that it make us notice things we would otherwise ignore. When we look at the grid as Warner has, we become aware of what enables the conditions of our living. We become aware, too, that we might *notice* things beyond pylons and power lines; we might notice humans and other animals, state actors and ecoystemic surrounds. Ironically, however, when we recognize the invisible ubiquity of the grid, we recognize, too, the poverty of our ethical conventions and habits, which rely so fully on the heroic humanism of private agency. And so a new type of agent arises as the hero of the environmental story: not the grid, but its governance. This, I think, is what Warner means when he says that supergriddedness is compatible with either fascism or democracy—that it is the maintenance of the grid that matters, the centralization of its network that must be secured.

Again, this is where desire is: It is pinned to power, not in the form of a natural resource but in the form of whatever means become necessary to making sure resources continue to exist and that they are used well, where "well" means effectively and still mostly without our noticing. Consider this remark from Agnes Martin: "When I first made a grid I happened to be thinking of the innocence of trees and then this grid came into my mind and I thought it represented innocence. And so I painted it and then I was satisfied. I thought, this is my vision."[28]

In 2009's *The Grid Book*, the art historian Hannah Higgins challenges both Krauss's association of the grid with modernism and its more general association with social control. The grid, she argues, is not just a sad suite of office cubicles, it is also the Guidonian hand of medieval musical notation, where "pitch values [are] attributed to the spaces between . . . lines as well as the lines themselves," so that the image of skin becomes an instruction for sound.[29] It's not just New York City's concrete net of avenues and streets, it is also the stick chart of the Marshall Islanders, who use palm fronds to weave waterproof charts that track wave patterns and currents. Grids, writes Higgins, "break down, shatter, bend, and adapt to unanticipated purposes, suggest[ing] that the homogenizing dimension of the grid-myth begs for reversal" or, you might say, for dialectic.[30] To that end, I'd like to return now to poetry, and to a poetics that exploits what Higgins characterizes as the insistent linearity of moveable type. For moveable type, too, is also a grid, famously blamed by Marshall McLuhan for having "translate[d] man from the magical world of the ear to the neutral visual world" of quiet cognition.[31]

This is the sort of cheerfully ahistorical formulation we expect from McLuhan, but in this case it also misses something important about literary engagements with griddedness. It is the invention of moveable type, after all, on which those elaborate Renaissance experiments with poetic pattern depend, as in the well-known example of

28. Agnes Martin, interview with Suzan Campbell, May 15, 1989, Archives of American Art, Smithsonian Institution, Washington, DC. Accessed at http://www.aaa.si.edu/collections/interviews/oral-history-interview-agnes-martin-13296. Last accessed June 18, 2022.

29. Hannah Higgins, *The Grid Book* (Cambridge, MA: MIT Press, 2009), 109.

30. Higgins, *The Grid Book*, 9.

31. Marshall McLuhan, *The Gutenberg Galaxy* (Toronto: University of Toronto Press, 2011), 26.

George Herbert's "Easter Wings," a poem printed in the shape of an angel's pinions.[32] Contra McLuhan, this poem is most assuredly lodged in a magical world of talismanic representation and sigils, where resemblance or sympathy between word and image transforms them jointly into an instrument of divine encounter. Where the warp of the text crosses over the weft of its shape—where the vertical line "combine[s]," to use Herbert's word, with the horizontal ones—Herbert's grid plan discovers a means not of social control, but of human elevation.

To be clear, these are grids of a very different sort than the ones Warner imagines, and it would be wrong to reduce "the grid" to a universal form independent of any specific material instantiation or practical function. Typography is not electrical engineering, and "Easter Wings" might save souls without saving the planet. Nor for that matter should we believe that a poem of this sort or any sort has the power to persuade us to adopt an environmental ethics, since environmental ethics has—as we've heard—so drastically misappraised its relationship to what individual actors do and think. A poem may raise our consciousness or refine our sensibility, but it won't convert the sun's rays into volts. As I said at the outset, my interest here lies in the moral intuitions art shapes and makes explicit, and it's entirely possible that moral intuitions are beside the point to the gridded future: power first, principles TBD. Let's imagine, then, an art, specifically a poetic art, that might accompany griddedness, that might be in a relation to it that is nonreferential but still diagnostic, descriptive, or discerning.

Consider Susan Howe's poem "Thorow," whose title reletters the last name of the author of *Walden* so that its homonymic affinities can be seen rather than heard: Thoreau, throw, thorough, even,

32. See the images at https://www.ccel.org/h/herbert/temple/Easterwings.html.

perhaps, tomorrow.[33] "Thorow" is also how Sir Humfrey Gilbert spells "through" when, in his 1576 essay *New Passage to Cataia,* quoted at the very beginning of Howe's book, he mourns the non-existence of a Northwest Passage *through* which the Pacific might be reached by means of the Arctic. "Thorow" is a poem about Henry David Thoreau, America, and what it means to inhabit the grid of fallen empires and failing economies. It is peppered with references, both direct and oblique, to imperial and counter-imperial massacres from Turner's Falls to Fort William Henry. At one point, Howe cites Thoreau's remark in a letter to a friend: "am glad to see that you have studied out the ponds, got the Indian names straightened out—which means made more crooked—&c. &c."[34]

In her own typography, Howe crosses the crooked with the straight, in a careful placement of phrases and spaces that results in something like the appearance of a decayed grid structure, through whose cables and nodes language none too smoothly passes. The lines realize or at least attempt a poetic historiography, one that records how a colonial enterprise of land-grabbing, resource extraction, and human and nonhuman displacement takes the shape of a plan whose aftereffects scatter and survive long after it has been executed. "Thorow" affirms the durability of technologies, from the technology of the settler to the technology of the book to the technology of lyric. It affirms, and it grieves, the durability of their consequences: the matter that doesn't decompose, the borders that stay in place, the forms that corner the market on cultural value. What would it take for these to become benign, if not, as Agnes Martin says, innocent? And what would it take—a different question—for them to change entirely?

33. Susan Howe, "Thorow," in *Singularities* (Hanover, NH: University Press of New England, 1990), 39. The volume can be viewed at https://archive.org/details/singularities0000howe. See, in particular, 56–58.

34. Howe, "Thorow," 42.

RESPONSE

Response to My Respondents

MICHAEL WARNER

It is every author's dream to have three responses so thoughtful, so accurate in their description of the lectures' arguments, so sympathetic to their aims, and so creative in mapping further conversations as the three in this volume.

Like all bits of good fortune, this one has a downside. Because the respondents have been so collaborative in spirit, I am deprived of the standard framing for the present assignment: the "reply to my critics." (Alas; that's always such fun.) A "response to my respondents" has nothing like the same drama. But I am grateful to each for reformulating the central questions instructively, and for showing how many questions remain about the grid's mediation of contemporary life. Each respondent has extended the historical narrative; each has insight into the antiethical nature of the grid; each has a sharp sense of the dilemmas raised for the politics of climate change.

In trying to think about the conditioning of our world by the grid, one major challenge is where to begin and end. For Jamieson, the problem of the grid is "an instance of a more general story about the stealth infrastructure, characteristic of modernity, that has substituted general, impersonal, anonymous, and structurally determined

On the Grid. Michael Warner, Oxford University Press. © Regents of the University of California 2025.
DOI: 10.1093/oso/9780197696248.003.0008

relationships for more particular, personal, face-to-face, episodic interactions."

> Instead of encouraging a sense of responsibility, these institutions of modernity often create episodes of guilt against a broad background of a sense of complicity that is not systematically connected to action. And so together we destroy the world but no one feels responsible. (p. 84)

For Nersessian, the telling context is "the very long age of Enlightenment" (p. 106). For Britton-Purdy, the scope widens to infrastructural dimensions of law and market, or to the long history of humans as "an infrastructure species" (p. 95). This plurality of framings seems to me not only sympathetic to my account but necessary for imagining how many lines of history converge on the present conjuncture.

At the same time, the lectures focus on a particularly modern kind of infrastructure because my concern has been how to think about energy. And energy—as a concept, as a problem for governance, as a matter of consumption—has only been around for a couple of centuries, and the grid form of energy only for about a hundred years. I focus on energy also because it introduces problems in our understanding of the environment beyond, say, the erosion of Gemeinschaft or the lengthening of supply chains that has been going on since the dawn of cities. Its symbolization as modern freedom, its assurance of capacities that are always ready and waiting for some kind of use, its regulation of our immediate environment as one of comfort and convenience—these features of grid mediation have come to control what we imagine to be acceptable as a green future.

My point has been that environmentalism's hopes, for better or worse, are being recalibrated to the realities of decarbonization. But is this in fact better? Or worse? If you thought that climate change would require a massive shift toward the green virtues of humility, mindfulness, cooperativeness, and respect for nature; or if you thought that it would uncouple capitalism from the demand for growth; or if you thought that alternative energy sources would allow distributed production and transparency to users; or if you believed any number of the other projected futures around which the critical efforts of environmental thought have clustered, then it might be time to notice that decarbonization efforts trend in a different direction altogether. In other words, worse.

But plenty of people think this is better. Even in the time since the lectures were delivered at Berkeley, I have seen an uptick in the number of people arguing that the real problem is not first-world excesses of consumption but rather the mentality of energy scarcity, of conservation, and even efficiency. The journalist Matthew Yglesias wants us to demand a world of "energy abundance": "We should raise our clean energy production ambitions. We don't want to replace 100% of our current dirty energy—we want to generate vastly more energy than we are currently using and make it zero carbon."[1] Ryan Avent similarly writes that "if you manage to disentangle growth in energy use from growth in the negative consequences of fossil-fuel use, then you are back to a world in which the harnessing of ever-greater amounts of energy per person is associated with transformative economic change and rising living standards. We run the risk of convincing ourselves that increasing the amount of energy we consume is bad *per se*, which is a very silly thing to believe when we could make everyone on the planet much, much better off by increasing the fraction of the sun's energy which we consume."[2]

1. Matthew Yglesias, "The Case for More Energy," *Slow Boring*, October 7, 2021, https://www.slowboring.com/p/energy-abundance.
2. Ryan Avent, "Was Coal the Low-Hanging Fruit? Or, the Promise of Abundant Clean Energy," *The Bellows*, July 8, 2021, https://ryanavent.substack.com/p/was-coal-the-low-hanging-fruit.

And J. Storrs Hall, aptly described by Ezra Klein as a "reactionary futurist," thinks that the last fifty years will be remembered not for the crisis of greenhouse gases, but for the stagnation in energy use caused in large part by the failure to adopt nuclear energy and "the rise of a counterculture hostile to progress."[3] Thinkers such as these tend to cite hydrogen fusion rather than nuclear power in their scenarios of illimitable capacity, but the dream they give voice to is clearly strong enough to keep looking for more power whenever finitude stands in its way.

Such rhetoric takes us right back to Etzler's paradise: costless power, exceeding all anticipated uses. This kind of thinking makes explicit some of the basic features of the grid that I have been discussing. As Heidegger perceived, grid-mediated energy moved the world beyond the order in which technology was a means to an end. It became the condition of availability for ends yet to be imagined. It ceased to be an object and became something more like environment in itself. I think this is roughly what Thoreau dreaded when he spoke of a transcendentalism in mechanics. Challenging nature in this way is as much a feature of decarbonization scenarios as it was in the heady days of nuclear power—as when Lewis Strauss, then chairman of the United States Atomic Energy Commission, predicted in 1954 a world in which "our children will enjoy in their homes electrical energy too cheap to meter."

And why not? In the lectures I tried to show that some of the central concerns of environmentalism—as in that list of green virtues such as mindfulness and humility—seem to lose traction in such responses to climate change; but one might say that the ideals of ethicalized awareness, collective agency, and democratic transparency

3. See https://press.stripe.com/where-is-my-flying-car.

are just archaic to a world marked, as Nersessian puts it, by "a seamless integration of the needs of the person with the load of the system" (p. 113). Why should something like ethicalized awareness be such a big deal?

Dale Jamieson's response in this volume is a helpful provocation here. He notes that I have written as though ethicalized awareness is "intrinsically valuable (or something like it)" (p. 85) and that one might have questions about such a norm. It is, for one thing, impossible to be aware of everything all the time. I would not go as far as Jamieson in saying that "attention is the scarcest of our resources" (p. 79) since attention can only be imagined as a *resource* or as *scarce* given the economization of attention that defines social media; nevertheless, he has a point.

Jamieson offers a thought experiment or counterfactual in which he imagines what might have happened if, when greenhouse gases were first identified as a problem, there had been a rational response in the corridors of power. Imagine that Exxon had not stonewalled the research and funded climate denialism, or that the Republican Party had not staked its future on fossil fuels, and that instead a consensus among technocrats and regulators had put saner policies quietly in place.

> On this scenario climate change would now be fading into the background toward invisibility, a moderately unpleasant problem that would occasionally poke its head into consciousness, but one that was more or less adequately being addressed, like ozone depletion. The arguments around climate change would have been nerdy, not the ones that we have today that oscillate between denial and moral and epistemological panic. This would be a better outcome than where we find ourselves in today, even though "ethicalized awareness" would not be central to the story. (p. 86–87)

Indeed, this would have been better than what we have today, though that is perhaps not saying much. Would we really have cause to lament the discreet opacity of such a technocratic solution? Don't we count on such solutions to most crises?

Perhaps a principal goal in the lectures is just to question the idea that climate change would be an opportunity to achieve some of the goals that twentieth-century environmentalism imagined, of which ethical awareness was an important part. And perhaps such goals could be consigned to the past of environmentalism. But Jedediah Britton-Purdy supplies a more telling answer to the question Jamieson raises. The counterfactual scenario, after all, is a good example of what Britton-Purdy calls a hack—a mode of intervention made possible by normal inattention; a solution that is not political at all, but fugitive and camouflaged; a heroism not of citizens but of a technical-administrative elite. Britton-Purdy's analogies with other modes of governmentality, such as the turn in transnational governance from citizens to stakeholders, or the economy of attention in the price system, are most helpful. These, he notes, are forms of antipolitics, and he has stated the problem so eloquently that I cannot do better than to quote it again: "It is a part of reflectiveness in modern life to be ambivalent about these forms of depoliticization, to be liberated by them from certain burdens and fears, and also to be haunted by awareness of falling into irresponsibility, of becoming the tools of our tools, and becoming other people's tools (or the tools of their tools), precisely as Thoreau warned" (p. 103).

Another objection to be made out here is that the grid, though conditioning us to normalized unconsciousness, does not prevent moments of awareness; thus one might imagine either future compromise formations (along the lines of "smart meters") or creative forms of politics and culture that bring new clarity and recognition to the grid-mediated world. This, I take it, is what Anahid Nersessian is working toward, when she helpfully imagines a counterpoetics of the grid. The rhetorical and representational forms of the grid can themselves be reimagined for purposes other than those of infrastructure. "Let's imagine, then, an art, specifically a poetic art, that might accompany griddedness, that might be in a relation to it that is nonreferential but still diagnostic, descriptive, or discerning" (p. 128). This seems like a fine kind of hope, and I am grateful to her for turning the otherwise skeptical tone of my analysis into something like a creative project. Running through all three responses is a recognition that the utopia of grid-mediated green energy challenges, but does not entirely foreclose, new kinds of collective deliberation and adaptation. These lectures were written in the hope of clarifying just such projects.

INDEX

For the benefit of digital users, indexed terms that span two pages (e.g., 52–53) may, on occasion, appear on only one of those pages.

INDEX